Journal of Neural Transmission
Supplementum 23

L. Oreland and B. A. Callingham (Eds.)

Monoamine Oxidase Enzymes

Review and Overview

Springer-Verlag Wien New York

Prof. Dr. Lars Oreland
Department of Biochemical Pharmacology, University of Uppsala, Sweden

Prof. Dr. Brian A. Callingham
Department of Pharmacology, University of Cambridge, Great Britain

With 28 Figures

ISSN 0303-6995
ISBN-13:978-3-211-81985-2   e-ISBN-13:978-3-7091-8901-6
DOI: 10.1007/978-3-7091-8901-6

Preface

The papers in this volume are the invited lectures at the Second Workshop on Amine Oxidases held at the Biomedical Center, Uppsala University, 2 to 4 August 1986.

It was a great pleasure for the organizers to see, in Uppsala, so many old and new friends, all brought together by our common interest in the fascinating field of Amine Oxidases. The Workshop was the successor of a long row of successful predecessors to start with the 1972 meeting in Cagliari, Sardinia, and recently preceded by the first meeting of Workshop design, held in Cambridge 1984.

Several of the previous meetings have been dedicated to distinguished senior research workers within the field. To start this Workshop a memorial lecture was given. This lecture, given by Prof. T. P. Singer, was dedicated to the late Prof. K. Kamijo, who was one of the pioneers of what is now worldwide interest in Amine Oxidases and who organized the superb meeting in Hakone 1981. To the great honour of all the participants the family of the late Prof. Kamijo visited the Workshop.

Abstracts of all the presentations have been published as a Supplement of Acta Pharmacologica et Toxicologica (Copenhagen).

For the success of the workshop we first wish to thank each author for his or her contribution. The invaluable support with respect to organization by Drs. M. Strolin-Benedetti and H. Kinemuchi made the Workshop possible. It is with great pleasure we acknowledge the dedicated support and readiness to improvise, whenever needed, by the local staff, Mrs. C. Rasmundson, Miss E. Danielsson, Miss G. Hallman, Drs. Eiko and Eiichi Sakurai, and Dr. S. S. Jossan.

At the end of the Workshop Dr. Franca Buffoni kindly sent us a preliminary invitation for the 3rd Workshop in Firenze 1988.

In conclusion we would like to thank Springer-Verlag in Vienna for the efficient and courteous way they have brought this Supplement to publication.

L. Oreland   **B. A. Callingham**

Contents

Singer, T. P.: Perspectives in MAO: past, present and future. A review 1

Tipton, K. F., O'Carroll, Anne-Marie, McCrodden, J. M.: The catalytic behaviour of monoamine oxidase . . . 25

Callingham, B. A., Barrand, M. A.: Some properties of semicarbazide-sensitive amine oxidases 37

Waldmeier, P. C.: Amine oxidases and their endogenous substrates (with special reference to monoamine oxidase and the brain) 55

Trevor, A. J., Singer, T. P., Ramsay, R. R., Castagnoli, Jr., N.: Processing of MPTP by monoamine oxidases: implications for molecular toxicology 73

Trendelenburg, U., Cassis, L., Grohmann, M., Langeloh, A.: The functional coupling of neuronal and extraneuronal transport with intracellular monoamine oxidase 91

Strolin Benedetti, Margherita, Dostert, P.: Overview of the present state of MAO inhibitors 103

Dowson, J. H.: MAO inhibitors in mental disease: their current status 121

Contents

J Neural Transm (1987) [Suppl] 23: 1—23

Perspectives in MAO: past, present, and future*

A review

T. P. Singer

Molecular Biology Division, Veterans Administration Medical Center, San Francisco, California, and Departments of Pharmaceutical Chemistry, Biochemistry and Biophysics, University of California, San Francisco, California, U.S.A.

Introduction

A plethora of international symposia have been devoted to monoamine oxidases in the past 15 years, including Cagliari, Sardinia (1971); CIBA Symposium, London (1975); Midland, Michigan (1979); Budapest (1979); Göteborg (1980); Airlie House, Virginia (1981); Hakone, Japan (1981); Mannheim, Germany (1982), and Paris (1983), and the First Workshop on MAO in Cambridge, England (1984). One might, in fact, wonder if progress in this field has been sufficiently rapid to justify the choice of MAO as the subject of these frequent tribal rituals. Probably not. Perhaps the reason lies elsewhere. Few topics in the bio-medical sciences interlink the interests of as varied a group of investigators in different disciplines as does MAO. One might add that even fewer problems require as much interdisciplinary effort for progress as does MAO, as has been dramatically illustrated in recent work on the experimental Parkinsonism induced by MPTP.

Although these frequent meetings have produced some excellent surveys of specialized aspects of MAO, such as its selective inhibitors or the potential for therapeutic use of these, general reviews of the subject have been few indeed (Blaschko, 1974; Singer, 1985 a). It seemed appropriate, therefore, to devote this lecture, dedicated to

* The original studies reported here were supported by Program Project HL 16251 from the NIH, by grant No. DMB 8416967 from the National Science Foundation and the Veterans Administration.

Professor Kamijo, to an overview of the subject. I shall attempt to summarize questions which appear to have been satisfactorily resolved in recent years or are near solution, then survey those which remain for ongoing or future investigations to answer.

Since space for this overview is limited, it cannot be comprehensive. The selection of topics covered is perforce somewhat subjective, guided, among other factors, by my areas of competence. It wouldn't do for a biochemist to evaluate conflicting clinical reports on the use of MAO inhibitors, for instance! I hope, therefore, that the omission of some references will not be mistaken for a value judgement on the work, although quite a few papers still appear in this field about which the less said, the better.

In his introduction to the Midland symposium on MAO, Von Korff (1979) presented a useful list of "unanswered questions". I thought that these might provide a suitable framework for the present overview. This is then my personal view of what progress has occurred in the intervening seven years in providing answers to these questions.

Inhibitors

Ever since the discovery in the early 1950s of the mood-elevating property of iproniazid and that this compound inhibited MAO, MAO inhibitors have been the bridge linking biochemical investigations with pharmacological and clinical studies on the enzyme. MAO inhibitors have been the subject of several recent reviews (Fowler and Ross, 1984; Singer, 1985 a, b), so that it may suffice here to summarize recent progress and questions which remain unanswered.

One of the questions posed by Von Korff (1979) was why irreversible ("suicide") inhibitors of MAO, e.g., clorgyline, deprenyl, trans-phenylcyclopropylamine (2-PCPA), and phenylhydrazine show such marked differences in their action on monoamine oxidases. With the growing concensus that MAO-A and B are different proteins (cf. below), and the demonstration (Singer, 1979; Singer and Salach, 1981) of the differences in the binding of A-selective, B-selective inhibitors to the two types of MAO, the different effect of clorgyline and deprenyl on either type of MAO is easily rationalized. Final proof may have to await elucidation of their tertiary structures by X-ray crystallography, which will then permit computer imaging of their substrate binding sites. Many of us would wager that the reason for the high affinity of clorgyline for the A enzyme and of deprenyl for the B type will then become obvious.

The reason for the differences in the action of acetylenic inhibitors (pargyline, clorgyline, and deprenyl), arylhydrazines, and cyclopropylamines on the same enzyme can also be rationalized in terms of the different nucleophiles with which the electrophilic form of the inhibitor generated by MAO combines in covalent linkage and the differences in partition ratios. The partition ratio—the proportion of the electrophilic form of the inhibitor generated which dissociates from the enzyme to that which combines irreversibly at the active site—varies widely even among the same types of compound. Thus, while the acetylenic amines, pargyline and deprenyl, can be used to titrate MAO-B preparations, since the reaction is rapid and stoichiometric (*i.e.*, the partition = 1), dimethylpropynylamine inactivates MAO-B more slowly and a large excess of the inhibitor is required for complete block (*i.e.*, partition ratio is high) (Maycock *et al.*, 1976). All acetylenic inhibitors of MAO-A and B react at the same site, however, forming a flavocyanine at N(5) of the flavin.

While the unique structure of the adduct formed with acetylenic amines has been firmly established since 1976 (Maycock *et al.*, 1976), the reaction of arylcyclopropylamines and arylhydrazines with MAO-A or B is more complex and may result in the formation of more than one type of adduct, depending on the exact structure of the inhibitor. Since bleaching of the 450 nm band of the flavoquinone accompanies the inactivation, it was thought by some investigators that 2-PCPA, as well as arylalkylamines (Silverman and Hoffman, 1979) alkylate the covalently bound flavin of MAO. It was shown, however (Paech *et al.*, 1979, 1980) that the flavin is not the source of the nucleophile which attacks the trans-phenylcyclopropylimine formed by the enzyme, but, rather a protein group, almost certainly the -SH group at the substrate binding site (Singer and Barron, 1945). This explains the reported slow reversal of the inactivation on dialysis against benzylamine (Hellerman and Erwin, 1968), as well as the fact that the radioactivity incorporated from 2-[2^{14}C]-PCPA during inactivation (1 mol/mol of inactive enzyme) is released on denaturation (Paech *et al.*, 1979, 1980), in accord with the known tendency of thiols to act as leaving groups. Silverman (1983) later showed that inactivation is accompanied by opening of the cyclopropane ring, so that the inactive enzyme yields 1 mol of cinnamaldehyde on denaturation with perchloric acid.

1-Phenylcyclopropylamine and 1-benzylcyclopropylamine behave similarly, but not not identically (Silverman and Zieske, 1985, 1986 a, b). The former reacts with two different nucleophiles in the ratio of 7 : 1, the substrate-site thiol, which slowly dissociates with time, and the reduced flavin, which forms an irreversible adduct.

The suicide inactivation of MAO-B by phenylhydrazine is even more complex, in part because the phenyldiazine formed in the initial dehydrogenation step has a partition ratio significantly greater than 1, showing that the majority is released as a product, in part because the fraction which forms a covalent adduct with the enzyme without dissociation is attacked by two types of nucleophiles, yielding, therefore, two different inactive forms. One nucleophile appears to be the C4(a) group of the reduced flavin, the other, most likely, the substrate-site -SH group (Kenney *et al.*, 1979). It was subsequently found (Singer and Husain, 1982) that while MAO-B partitions equally between these two types of nucleophiles, with the A enzyme ~ 70% of the label from ring-labeled phenylhydrazine is recovered in the protein, and 30% in the flavin peptide. The distribution between the two types of products is also a function of the length of the alkyl chain.

Of interest is also the reaction of β, γ, δ-allenic amines with MAO. Several years ago (Krantz *et al.*, 1979) it was reported that these amines, unlike the isomeric acetylenic amines, do not form a flavocyanine at N(5) during the irreversible inactivation of MAO-B but may be forming, instead, an unstable cyclic C4(a)—N(5) adduct. More recent studies in Krantz's laboratory (White *et al.*, 1983) pointed to the importance of chirality on the inactivation of the B enzyme by a series of allenic amines.

It is clear from this summary that even with irreversible (or pseudoirreversible) inhibitors the type of nucleophile which attacks the electrophilic compound formed from the inhibitor depends on the structure and cannot be a priori predicted. In case of the reversible inhibitors of MAO-A or B, it is even more difficult to predict or to demonstrate the reaction site, since with these compounds no stable adduct forms, which could be isolated and characterized.

A significant, recent development in this field has been the unexpected demonstration in our joint study with Drs. Castagnoli and Trevor that MPTP (1-methyl-4-phenyl-1, 2, 3, 6-tetrahydropyridine) and its oxidation products MPDP$^+$ (the dihydropyridinium form) and MPP$^+$ (the pyridinium form) are excellent reversible, competitive inhibitors of both MAO-A and B (Salach *et al.*, 1984; Singer *et al.*, 1985, 1986). Further, we have shown that MPTP and MPDP$^+$ are also good mechanism-based (*i.e.*, suicide) inhibitors of MAO-B and A. I shall not dwell on these, as Dr. Trevor will discuss the subject in detail, except for one aspect, which is relevant to the discussion above.

The inactivation of MAO-B by MPTP, like that elicited by phenylhydrazine, seems to involve the formation of more than one

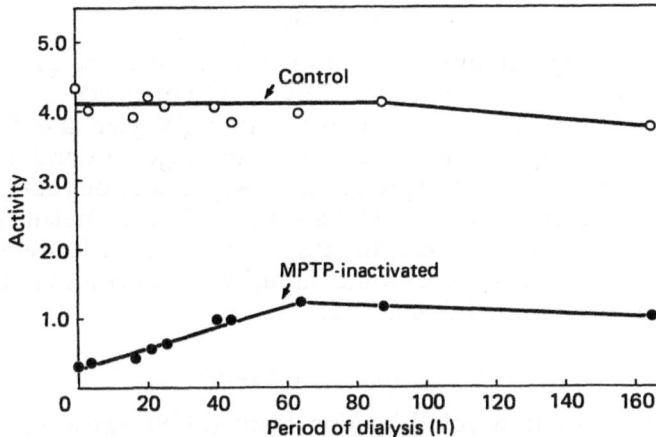

Fig. 1. Partial reactivation of MPTP-inactivated MAO-B by dialysis. A sample of the enzyme was inactivated with 5 mM-MPTP at 30 °C and uncombined MPTP was removed. Aliquots (1 ml each) were then dialyzed against phosphate buffer, pH 7.2, containing 2 mg of Triton X-100/ml at 0 °C. Samples were removed and assayed with 3.3 mM-benzylamine at the times shown. The control was similarly treated, except for preincubation with MPTP. (From Singer *et al.*, 1986)

adduct with the enzyme, because on extended dialysis a fraction of the catalytic activity returns (Fig. 1). This suggests that the loss of activity is the result of an irreversible and of a (slowly) reversible reaction, as has been demonstrated for 1-phenylcyclopropylamine and MAO. The reversible reaction has been suggested to involve the thiol group, forming a thiohemiketal with an oxidation product of MPTP (Singer *et al.*, 1985). This is also in accord with the observation that after proteolysis of the inactivated enzyme considerable amount of the flavin peptide has been recovered with the original flavoquinone spectrum. This would not occur, of course, if all the inhibitor were covalently linked to a nucleophilic group on the reduced flavin.

Questions concerning single forms of MAO

Flavin component of MAO-A

The first question raised by Von Korff (1979) concerning the individual enzymes was whether the same type of covalently bound flavin was present in the A and B enzymes. This question was answered several years ago by the demonstration (Nagy and Salach, 1981) that both the 8α-S-cysteinyl linkage and the peptide sequence near the flavin were identical in the A and B enzymes.

Role of iron

Another long-standing assumption that Fe may be present in MAO-B has been laid to rest by removing all but traces of Fe from purified MAO-B by gradient centrifugation (Weyler and Salach, 1981). The resulting enzyme appears to be homogenous and no longer displays the 412 nm band present in most preparations but is fully active, showing that Erwin and Hellerman (1967) and Oreland (1971) were correct in predicting that this absorbance and the Fe were associated with a tenacious cytochrome impurity. Iron has also not been detected in the purified A enzyme.

The question of subunits

The next question posed by Von Korff (1979) was whether the subunits of MAO-B are "truly identical" and how they compare with the subunits of MAO-A. Despite considerable effort in many laboratories, to this day we can only supply a partial answer to these important questions.

The difficulty in arriving at reliable answers to these questions hinges on the dual facts that (1) MAO preparations tend to aggregate and bind avidly large amounts of detergents, rendering physical methods for molecular weight determination of questionable value and (2) the circumstance that the same type of MAO, *e.g.,* the B enzyme, may differ in subunit molecular weight by as much as $2,000 \pm 500$ in different mammalian livers (Weyler and Salach, 1985). In fact, reliable values for the native molecular weight of the A enzyme are not yet available. For the B enzyme a native molecular weight of 110 to 120 kilodaltons in different species (beef kidney and liver, pig liver) seems to be a reasonable value, based on pargyline binding or flavin content per biuret or Lowry protein and, in one case, determined by gel exclusion in the presence of cholate (Chuang *et al.,* 1974; Oreland, 1971; Salach, 1979). Values for the subunit molecular weight of the A and B enzyme from the literature are summarized in Table 1. While the differences between the A and B enzymes in the Table seem insignificant, considering the variation in the value for the B enzyme in different species or cell types, or even in the same tissue as determined in different laboratories, nevertheless, it may be concluded that the subunits of the A enzyme are somewhat larger than those of MAO-B, when the two are subjected to gel electrophoresis side-by-side. The difference may be of the order of 1,000 daltons (Weyler and Salach, 1985).

Since 115 K seems to be a reasonable value for the native molecular weight of MAO-B but only a single subunit has been found in gel

Table 1. Estimates of the subunit molecular weights of MAO-A and B
determined by SDS-Page

Tissue source	MAO-A	MAO-B	reference
Human platelet		59,000	Denney *et al.*, 1982
Human platelet		60,000	Cawthon *et al.*, 1981
Human placenta	67,000		Brown *et al.*, 1980
Human placenta	63,000		Cawthon *et al.*, 1981
Human placenta	64,000		Kochersperger *et al.*, 1985
Human placenta	62,000–64,000		Weyler and Salech, 1985
Human fibroblast	63,000	60,000	Cawthon *et al.*, 1981
Rat hepatoma MH_1C_1	63,000	60,000	Cawthon and Breakefield, 1983
Rat glioma C_6	63,000		Ibid.
Rat liver	60,000	55,000	Callingham and Parkinson, 1979
Bovine liver		61,000	Salach, 1979
Bovine liver		52,000	Minamiura and Yasunobu, 1978

electrophoresis, it has been concluded that the enzyme contains two
similar sized subunits, which differ in at least one regard, that only
one of them contains covalently bound flavin. As noted above, this
conclusion cannot as yet be drawn for the A enzyme.

Function of phospholipids

It is difficult to trace the origins of the notion that phospholipids
are essential for the activity of MAO and that, by extrapolation, differences between MAO-A and B are due to lipids being attached to
the same protein. Perhaps it started with the observation of Houslay
and Tipton (1973) that the treatment of rat liver preparations with
strong perchlorate reduces the differences between the A and B
forms. It was reasoned that, since perchlorate weakens non-covalent
interactions, such as those which may bind proteins to lipids, the
results may be taken to indicate that the differences in specificity between the A and B enzymes are due to the particular lipids attached.
The same interpretation was ascribed by some workers to Oreland's
finding (1971) that the extraction of pig liver mitochondria with
methyl ethyl ketone leads to loss of the A activity but recovery of the
B enzyme. However, perchlorate, as well as methyl ethyl ketone,
inactivate and denature proteins. Thus, when a source containing
both forms of MAO is treated with either reagent, the A activity is
more apt to disappear, since MAO-A is more sensitive to denaturation. This fact was, in fact, eloquently pointed out by Ekstedt and
Oreland (1976) but is not taken into account by adherents of the
"delipidation" theory.

In recent years the two main proponents of the notion that the differences between the A and B enzymes arise from different phospholipid environments have been the laboratories of Huang and of Yagi. Both base their conclusions on delipidation and reinsertion into liposomes. It is important to understand the experimental techniques used, for there lies the reason for the erroneous conclusions.

In the experiments of Huang and Faulkner (1981) phospholipase A_2 was used to "inactivate" the A and B enzymes in brain preparations, a fairly selective procedure, which lowers the phospholipid content of preparations markedly without denaturing proteins. Their assay procedure was a non-kinetic method of limited accuracy, and thus could not be used to characterize the sequence of events. It should be noted that MAO-B from both beef kidney (Erwin and Hellerman, 1967) and liver (Salach, 1979) and MAO-A from human placenta (Weyler and Salach, 1985) have been reported not to lose activity on digestion with phospholipase A_2, but in these studies initial rate measurements and reliable assays were used. The studies of Yagi (cf., *inter alia*, Naoi and Yagi, 1980) are beset with problems as regards the delipidation method, the assay, and even the identity of the enzyme studied. Delipidation was brought about with either SDS and dithiothreitol or with sodium perchlorate, both of which are unfolding agents and they should not be used for delipidation. The assays were not suited for distinguishing effects on K_m and V_{max} and the identity of the enzyme used in the study is puzzling. It fits none of the known properties of mitochondrial MAO-A or B in terms of substrate specificity or subunit molecular weight (28 kilodaltons) and is of exceedingly low activity. In view of these shortcomings, the conclusion that the lipid environment regulates the substrate specificity of MAO seems unjustified.

One should also take into account that MAO-B from beef kidney (Erwin and Hellerman, 1967) contains no lipids and that Salach's preparation from liver may be delipidated nearly completely with little loss of activity at V_{max} (Husain *et al.*, 1981). Further, Zeller has often noted in his lectures that his preparation of MAO-A from human placenta is also lipid-free but highly active.

Recognizing the greater sensitivity of MAO-A than of B to inactivation by a variety of agents, the results of investigators who used chaotropic or unfolding agents for delipidation may be rationalized. The observations of Huang's laboratory that phospholipase A_2 treatment seems to lower the activity of brain mitochondria on both A and B substrates and that reinsertion into liposomes increases the activity way beyond the level in the intact mitochondria cannot be explained this way. A thoughful paper by Navarro-Welch and

McCauley (1982), however, has helped to reconcile their report with observations in several other laboratories that similar treatment does not lower the catalytic activity of MAO from other mammalian sources. Using similar procedures for delipidation and reconstitution as Huang and Faulkner (1981) but rat liver mitochondria, Navarro-Welch and McCauley have shown that depletion of phosphatides and reconstitution with phospholipids do not alter either A or B activity at V_{max} but that phospholipids lower the K_m values of both enzymes for their amine substrates, i.e., increase the apparent affinity. Moreover, phospholipids in the incubation medium stimulate both A and B activity in intact, untreated mitochondria. It is concluded that phospholipids cannot be regarded as essential components of the enzyme, nor the effect of added phospholipids to a delipidated sample as a reconstitution, and the apparent stimulation is due to unincorporated phospholipids in the medium.

My personal view is very close to this. Monoamine oxidases function in a lipid environment in the membrane in the cell. As is true of several other membrane-bound enzymes, the activity of MAO-A and B may well be influenced by the lipid environment. Whether this is the result of greater accessibility of substrates containing hydrophobic groups to the active site or of different conformations the enzymes assume in membranes of varying fluidity is not clear but is possible to investigate experimentally. The point to remember, however, is that phospholipids should no longer be regarded as essential and integral components of MAO.

Function of -SH groups

It has been known for over 40 years (Singer and Barron, 1945) that at least one -SH group is required for the activity of MAO-B from liver. Over the years numerous papers dealt with the subject, the majority confirming that an -SH group was involved in the activity, with one report (Gomes et al., 1969) claiming that cysteine was not present at the active site. The same group (Gomes et al., 1976) later reported on the basis of titrations with DTNB and mercurials that there are two essential active site cysteines. In more recent studies in our laboratory, using DTNB or the structurally related 2, 2'-dipyridyl disulfide for titrations (Weyler and Salach, 1985), we also noted that two equivalents of the -SH reagent were required to abolish the activity of beef liver MAO-B. Titration of the native enzyme with CH_3HgI, iodoacetamide, or NEM in both Yasunobu's laboratory (Yasunobu et al., 1979) and ours (Paech and Singer, 1983) showed very different reaction rates and different number of cysteine residues, and no indication that the substrate site thiol could be differentially alkylated with these reagents. There is even some discrepancy in the total number of half-cysteine residues detected by chemical analysis of the denatured protein: Minamiura and Yasunobu (1978), found 13 to 16 per mol of flavin, while our figure is close to 12.

These uncertainties do not pose an impediment to progress in understanding the action of MAO-A and B *in vitro* and *in vivo*, though they will have to be dealt with when crystallographic data of their structures become available. What matters at the moment and what Von Korff (1979) may have had in mind in his query about "the role of -SH groups" in the enzyme is the paradox posed by the finding that two thiol groups have to be chemically modified for complete inactivation of MAO-B, although only a single substrate binding site seems to be present. The latter became apparent when it was demonstrated (Paech *et al.*, 1979) that 1 mol of 2-phenylcyclopro-pylamine is required to inactivate 1 mol of MAO-B, that the inactivation and incorporation of radioactivity from [^{14}C]-2-PCPA into the protein go hand in hand, and that inactivation involves almost certainly the substrate-site thiol. This conclusion has been reinforced by the report (Silverman and Zieske, 1986) that the covalent but reversible binding of 1-[phenyl-^{14}C]PCPA to MAO-B is to a substrate site -SH group. Thus, titration of the cysteine residues of the denatured enzyme with DTNB gave 6.2 cysteine residues in controls, but only 5.2 residues in samples pretreated with 1-PCPA. We have confirmed this and found 8 residues in the absence and 7 in the presence of 1-PCPA.

The resolution of this paradox may be that DTNB and similar reagents react at comparable rates with 2 cysteine residues in the enzyme, only one of which is at the active site. Since the statistical probability of an "essential" and of a "non-essential" -SH being oxidized is equal, it will appear that 2 thiols must react before inactivation is complete. In contrast, the suicide inhibitors 2-PCPA and 1-PCPA yield the reactive electrophilic product while bound to the active site and will be immediately attacked by adjacent nucleophiles, so that there is no chance for a reaction with a second thiol which is not in close proximity.

Substrate specificity

At the time of the Lansing meeting in 1979, the substrate specificities of MAO-A and B did not seem to be "unanswered questions"

(Von Korff, 1979). On the contrary, Zeller (1979) offered a systematic classification of amine oxidases, based on their known substrate specificities. The discovery (Salach *et al.*, 1984) that MPTP, a tertiary amine, is not only oxidized by MAO-B but is one of its best substrates, both in terms of V_{max} and K_m values, and that MPTP is also reasonably fast oxidized by MAO-A was unexpected. Tertiary amines, such as pargyline, deprenyl, and clorgyline, had long been known to be oxidized by MAO, albeit slowly (otherwise they could not be suicide inhibitors of the enzyme), but this was, as far as we know, the first instance of the rapid processing of a tertiary amine. Moreover, 2, 3-MPDP⁺, the dihydropyridinium species arising, is also oxidized by MAO-B, although the rates are not nearly as high as with MPTP (Singer *et al.*, 1986). Though there are statements in the recent literature that MPDP⁺ is not a substrate of MAO, this is clearly wrong, since it is a suicide inhibitor of the enzyme (Singer *et al.*, 1985, 1986).

The interaction of MAO-A and B with MPTP and its oxidation products is more fully discussed by Dr. Trevor in his volume. My purpose in calling attention to these studies is that they point to a need to explore the substrate specificities of the two forms of MAO further, particularly as regards their action on tertiary amines. Studies along these lines are in progress, in fact, in several laboratories. The results so far are summarized in Fig. 2 and are more fully discussed in Dr. Trevor's paper.

MPTP : R = CH₃

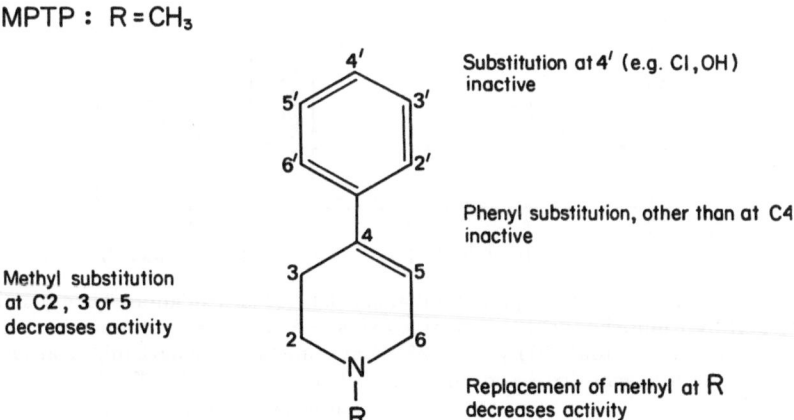

Substitution at 4′ (e.g. Cl, OH) inactive

Phenyl substitution, other than at C4 inactive

Methyl substitution at C2, 3 or 5 decreases activity

Replacement of methyl at R decreases activity

Fig. 2. Phenyl tetrahydropyridines as substrates of MAO-B

Mechanism

During the past few years significant progress has been made in understanding the catalytic mechanism of action of MAO-B. Until recently it was assumed that the enzyme acts by a ping-pong mechanism because of the parallel-line kinetics obtained in a number of laboratories for the enzyme from various species. Bright and Porter (1975), however, pointed out that such steady-state kinetic data are insufficient to draw conclusions as regards the kinetic mechanism of an enzyme, unless supported by pre-steady state data. Husain *et al.* (1982), therefore, investigated in detail the kinetic mechanism of beef liver MAO by pre-steady state and steady state methods, including the measurement of kinetic isotope effects. With benzylamine as substrate a very large primary kinetic isotope effect was found, showing that the initial proton transfer to the flavin is rate-limiting. Steady-state experiments on the reoxidation of the benzylamine-reduced enzyme by O_2 showed only a moderate dependence on O_2 concentration (Fig. 3). In contrast, the reaction of phenylethylamine with MAO showed almost no kinetic isotope effect in steady state assays but a very large dependence of the oxidative half-reaction on O_2 concentration (Fig. 4), showing that with this substrate reoxidation of MAO by O_2 is rate-limiting. In fact (Table 2), the rate of reduction of MAO by β-phenylethylamine is huge (34,300 min^{-1}),

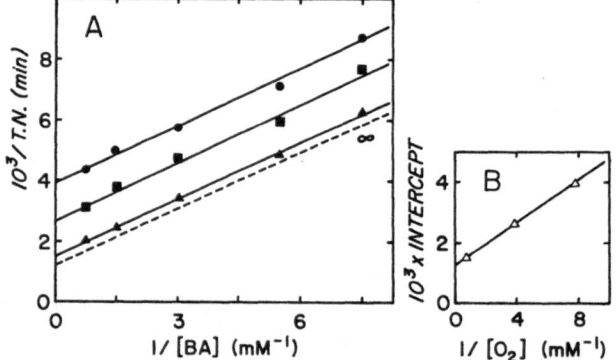

Fig. 3. **A** Double-reciprocal plot of initial rate data for the oxidation of benzylamine by beef liver mitochondrial monoamine oxidase. Assays were performed in 50 mM Hepes and 0.5% Triton X-100, pH 7.5, 25 °C, as a function of benzylamine concentration at the following fixed concentrations of oxygen (●) 0.13; (■) 0.26; and (▲) 1.3 mM. The initial rates were measured by following the increase in absorbance at 250 nm. **B** Secondary plot of intercepts from the data in (A) against the reciprocals of the oxygen concentration. BA denotes benzylamine. (From Husain *et al.*, 1981)

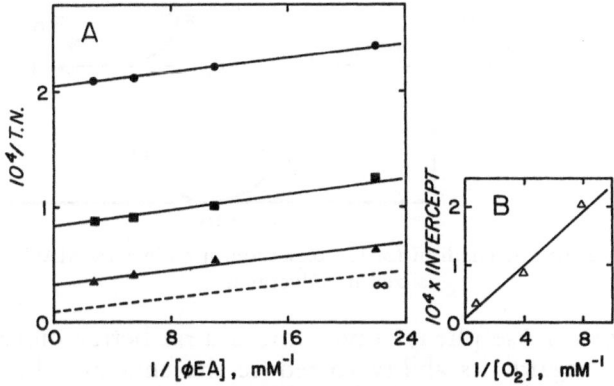

Fig. 4. Steady-state analysis of the oxidation of phenylethylamine ($\emptyset$ EA). Conditions as in Fig. 3

Table 2. Kinetic constants calculated from the reductive half-reaction data

	Values			
Constant	Benzyl-amine	α, α-$^2H_2$]-Benzyl-amine	β-Phenyl-ethylamine	β-[α-α-$^2H_2$]-Phenyl-ethylamine
k_3 (min^{-1})	700	80	34,300	11,290
K_S (k_2/k_1) (mM)	0.36	0.33	4.5	3.3

compared with the catalytic turnover in steady state assays (1,250 min^{-1}). The conclusion from this work is that the kinetic mechanism of MAO-B depends on the substrate used. With phenylethylamine a binary complex (ping-pong) mechanism operates (upper loop of Scheme 1), while in the case of benzylamine (lower loop of Scheme 1) a tertiary complex is involved. It is interesting to

Scheme 1

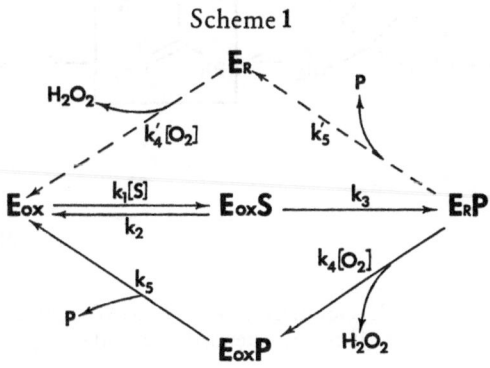

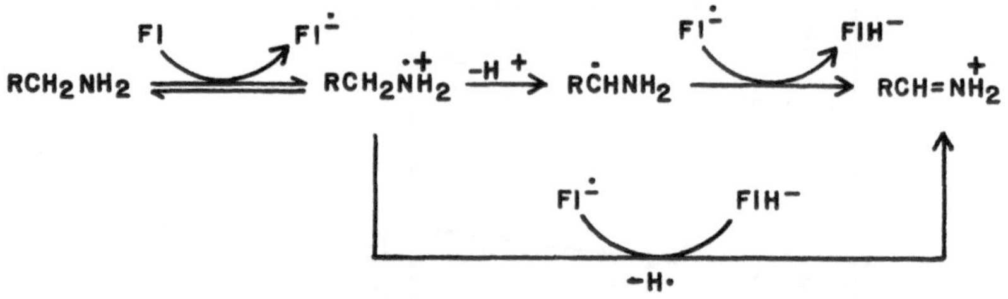

Fig. 5. One electron mechanism for the oxidation of amines by MAO-B, as visualized by Silverman

note that in a sense phenylethylamine is a far better substrate for MAO-B, as regards its ability to reduce the enzyme, although at ambient O_2 concentrations in the cell the overall rate of oxidation is far less than that of benzylamine.

These results were confirmed by Pearce and Roth (1985) using crude mitochondria from human brain and clorgyline to block the A enzyme. They also demonstrated that tyramine and tryptamine, like phenylethylamine, are deaminated by a binary complex mechanism. Similar studies on MAO-A have not been reported, probably because of the limited amount of enzyme which has been available from human placenta.

Three possible mechanisms for the initial events in the action of MAO on amines were suggested by Maycock et al. (1976). Of these, the carbanion mechanism was ruled out several years ago, in part on the basis of comparison of isomeric allenic and acetylenic amines on MAO-B (Krantz et al., 1979). No evidence has been found for a carbonium mechanism but a radical mechanism was proposed by Silverman et al. (1980) and Krantz et al. (1979), which has been slightly

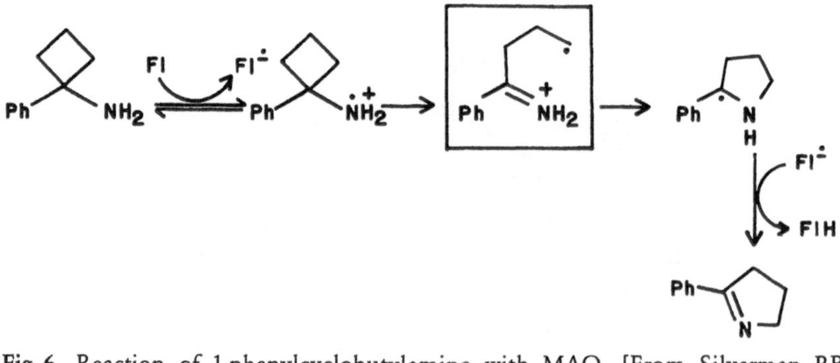

Fig. 6. Reaction of 1-phenylcyclobutylamine with MAO. [From Silverman RB, Zieske PA (1980) Biochemistry 25: 341–346]

revised by Silverman's group over the years to the form shown in Fig. 5, on the basis of extensive studies with cyclopropylamine analogues.

Although so far no flavin radical has been detected during the action of MAO, this may be due to the possibility that the initial rate-limiting one electron transfer is rapidly followed by a second, faster electron transfer. The radical mechanism is strongly supported by studies on the reaction of 1-phenylcyclobutylamine with MAO-B (Silverman and Zieske, 1986). The product isolated from this reaction is that expected for radical cyclization of the proposed intermediate (Fig. 6).

Biosynthesis

The recovery of MAO-A and B activities in rat organs following the administration of large amounts of the appropriate irreversible inhibitor (clorgyline, deprenyl, or pargyline) has been studied by several investigators over the years, notably in Kamijo's laboratory (Goridis and Neff, 1971; Erwin and Dietrich, 1971; Egashira and Kamijo, 1979; Egashira and Yamanaka, 1981). The latter group has studied the recovery of activity on serotonin and β-phenylethylamine in both liver mitochondria and microsomes and found the recovery to be much faster in the microsomal fraction. An unexpected finding in these studies was that MAO-A and B activities were present in both the mitochondrial and microsomal fractions at significant levels (Egashira and Yamanaka, 1981). Mitochondrial contamination in the microsomes was controlled by assaying for succinate dehydrogenase, which is, however, an inner membrane enzyme, so that sloughing off of the outer membrane during isolation—which would result in recovery of outer membrane material in the microsomes—would not have been detected. This would not account, however, for these authors' finding of a much faster recovery of MAO-A activity in the "microsomal" than in the "mitochondrial" fraction in livers form clorgyline-treated rats.

The explanation offered by these authors is that the activity in the microsomes represents newly synthesized enzyme, a precursor of mitochondrial MAO, not yet inserted into mitochondria. If so, the finding of a significant fraction of the total MAO activity in the microsomes even in the control animals is surprising, particularly since other authors have not found major MAO activity outside the outer membrane, otherwise it could not be used as a marker enzyme. Their interpretation also implies that the transport of MAO into mitochondria is a very slow process compared with its synthesis and that the enzyme is fully assembled outside of the mitochondria, including incorporation of the cysteinyl flavin prosthetic group—otherwise it would not have been detected in activity measurements. These puzzling findings could be perhaps explained by subjecting the "microsomal" fraction to gradient centrifugation and comparing the distribution of MAO and of glucose-6-phosphatase in the resulting fractions.

These studies, unfortunately, have not shed any light on the interesting question whether the covalently bound flavin is inserted cotranslationally or post-translationally and by what mechanism.

Multiple forms of MAO

The interrelations of MAO-A and B has been a favorite subject of investigations, theories, speculations, and even heated debates in this field for nearly two decades. The three ideas which have been advanced—that MAO-A and B are the same protein with different lipids attached, that they are a single protein with two different active sites (an "A" site and a "B" site), and that they are different proteins, individually coded, have been examined in several recent reviews (Singer, 1985 a, c), so that a summary of the current status of the problem may suffice.

In the foregoing section I discussed why it no longer seems logical to continue assuming that lipids are essential for the activity of either form or that different lipids impart differences in substrate specificity to the same protein. The notion of a single protein with two different substrate sites appears to us also to be untenable in the light of the facts that it takes one mol of 2-PCPA or of deprenyl, clorgyline, or pargyline, all of which are mechanism based inhibitors processed at the active site, to inactivate one mol of MAO-A or B, resulting in the binding of one mol of radioactive adduct formed from the inhibitor. This is clearly incompatible with the notion of two substrate sites on the same protein. Moreover, since there is a single mol of cysteinyl flavin in both MAO-A and B, if two substrate sites were present feeding electrons to the flavin, then the A and B binding sites would compete for the flavin. Given the fact that with many substrates the initial proton abstraction is rate-limiting, it follows that A and B substrates should be mutually inhibitory in any MAO preparation. This is clearly not the case. Finally, how would one explain then that some tissues have only MAO-A, others only MAO-B present?

Strong evidence that MAO-A and B are different proteins has been steadily accumulating in recent years. One line of evidence involves comparison of nearly homogenous A and B. Their molecular properties—such as stability, water solubility, etc., are quite different. No one who has worked with highly purified preparations of both can escape the impression that they are quite different proteins. Moreover, as already discussed, their subunit molecular weights are consistently found to be different. The second line of evidence has come from the finding of different peptide patterns following labeling with [^{3}H] pargyline and proteolysis of the A and B enzymes (Cawthorn and Breakfield, 1979; Cawthorn et al., 1981). In this study the A enzyme was selectively labeled by preincubation of the preparations with deprenyl and the B enzyme by preincubation with clorgyline, both of which, of course, bind at the same site as pargyline. In order to preclude the possible criticism that the differences noted were the results of the effect of different post-translational modifications of the same protein on the specificity of proteolytic cleavage, Smith et al. (1985) included cyanogen bromide cleavage in the procedure and again found major differences in the peptide maps of MAO-A and B. Possibly, the only part of the two proteins which has for certain the same amino acid sequence is the region surrounding the flavin (Nagy, J., and Salach, J. I., 1981), which is, of course, the domaine most apt to be conserved.

The body of immunochemical evidence that MAO-A and B are different proteins has been growing since the report of McCauley and Racker (1973) that the two enzymes can be distinguished by immunochemical means. The next important step was the development of polyclonal antibodies to pure bovine liver MAO-B (Levitt et al., 1982; Pintar et al., 1983) and the demonstration that they do not cross-react with MAO-A from a variety of sources. Monoclonal antibodies to human MAO-B have also been raised, using the partially purified enzyme from platelets (Denney et al., 1982) and have been used to separate MAO-A and B from human liver on immunoaffinity columns. The most recent advance has been the development of monoclonal antibodies to human MAO-A (Kochersperger et al., 1985) and the demonstration that they do not immunoprecipate MAO-B from several sources, including human platelets.

Although the evidence favoring two distinct proteins is impressive, final proof may have to await the demonstration of different amino acid sequences in the two enzymes. Despite the relative scarcity of pure MAO-A, the time is ripe to compare the sequences at the substrate binding sites, which must be different in the two enzymes in order to account for differences in their substrate and inhibitor specificities. A study along these lines is, in fact, in progress.

Even after the non-identity of the two forms is conclusively established, some of Von Korff's (1979) questions concerning them will remain to be answered. Questions such as the reason for the markedly different A : B ratios in the same organ of different species or in different tissues of the same species remain as puzzling as they were in 1979.

Plasma amine oxidase

For many years the identity of the prosthetic group of plasma amine oxidase eluded identification. It was generally agreed that copper is an essential component, but the covalently bound organic cofactor remained unknown. The only clue to its identity was that carbonyl reagents inhibited the enzyme. This and the absorption spectrum had lead to the erroneous conclusion that pyridoxal phosphate is involved, but analysis failed to detect the presence of B_6. Recently, two laboratories independently reported that the elusive cofactor is pyrroloquinoline quinone (PQQ, also called methoxatin), a coenzyme previously detected only in bacterial dehydrogenases, notably methylotrophs. Lobenstein et al. (1984) isolated the 2, 4-dinitrophenylhydrazone of PQQ by proteolysis and chromatography in very low yield and identified it by HPLC, H-NMR, and electronic spectrum. What is not clear from this brief report is how come the hydrazone of free PQQ was obtained, albeit in a very small yield, when the cofactor is covalently bound to the peptide chain in the enzyme. The interesting question of the site of attachment of the terminal amino acid to the PQQ ring system and the nature of the linkage remain wide open.

Ameyama et al. (1984), on the other hand, used proteolytic digestion, followed by hydrolysis in 6 N HCl at 110 °, to obtain free PQQ, which they identified by electronic and fluorescence spectra and enzyme assay. They also found PQQ present in diamine oxidase from pig kidney, showing that both nonflavin amine oxidases in animal tissues contain apparently PQQ as prosthetic groups. These two brief reports may open the door to systematic investigations of the possible presence of PQQ in other eukaryotic enzymes, of the type of covalent linkage, and its formation during biosynthesis and of the mechanism of action of plasma amine oxidase and diamine oxidase. The situation may be comparable to the field of covalently bound flavins in 1954, when the first representative of this important group of coenzymes was discovered in purified succinate dehydrogenase.

Concluding remarks

The foregoing overview has not touched on the second set of unanswered questions posed by Von Korff (1979): those of a physiological, pharmacological, and clinical nature. I have avoided surveying progress in these areas, since there are many participants in this workshop better qualified than I am to do so. Fortunately, the limitations of space provide a sound excuse for this omission. I would be remiss, however, if I did not point out that basic biochemical investigations on MAO continue to provide the tools and sometimes the motivation and the clues for progress in the pharmacology and clinical applications of MAO inhibitors and have even provided clues to the etiology of neurological disorders. The story of the biochemical reactions which lead from the bioactivation of MPTP by MAO to nigrostriatal degeneration is a fascinating example and will be surveyed by Trevor.

The conversion of MPTP, by itself an innocuous substance, to highly neurotoxic forms by MAO may also serve as a reminder that this enzyme not only detoxifies some amines but can create toxic products from other xenobiotics.

The isolation of pure MAO-A and B paved the way to the development of antibodies, which have already been responsible for some fascinating findings concerning the sympathetic nervous system. Thus, Levitt et al. (1981) and Westlund et al. (1985) have published evidence showing that MAO-B, but not A, is present in serotonergic neurons, while only the A form was detected in dopamine cells, suggesting that these enzymes protect neurons by degrading extraneous amines and false neurotransmitters. Lastly, the unexpected announcement of Knoll (1981) of the dramatic effect of deprenyl on the sexual activity of old and sluggish rats has important implications on the quality of life in senescence.

References

Ameyama M, Hayashi M, Matsushita K, Shinagawa E, Adachi O (1984) Microbial production of pyrolloquinoline quinone. Agric Biol Chem 48: 561–565

Blaschko H (1974) The natural history of amine oxidases. Rev Physiol Biochem Pharmacol 70: 84–148

Bright HJ, Porter DJT (1975) Flavoprotein oxidases. In: Boyer PD (ed) The enzymes, vol 12, 3rd edn. Academic Press, New York, pp 421–505

Brown GK, Powell JF, Craig IW (1980) Molecular weight differences between human platelet and placental monoamine oxidase. Biochem Pharmacol 29: 2595—2603

Callingham BA, Parkinson D (1979) Tritiated pargyline binding to rat liver mitochondrial MAO. In: Singer TP, Von Korff RW, Murphy DL (eds) Monoamine oxidase: structure, function, and altered functions. Academic Press, New York, pp 81—86

Cawthorn RM, Pintar JE, Haseltine FP, Breakefield XO (1981) Differences in the structure of A and B forms of human monoamine oxidase. J Neurochem 37: 363—372

Cawthorn RM, Breakefield XO (1983) Differences in the structure of monoamine oxidases A and B in rat clonal cell lines. Biochem Pharmacol 32: 441—448

Chuang HYK, Patek DR, Hellerman (1974) Mitochondrial monoamine oxidase. Inactivation by pargyline. Adduct formation. J Biol Chem 249: 2381—2384

Denney RM, Fritz RR, Patel NT, Abell CW (1982) Human liver MAO-A and MAO-B separated by immunoaffinity chromatography with MAO-B specific monoclonal antibody. Science 215: 1400—1403

Egashira T, Kamijo K (1979) Synthetic rates of monoamine oxidase in rat liver after clorgyline or deprenyl administration. Japan J Pharmacol 29: 677—679

Egashira T, Yamanaka Y (1981) Further studies on the synthesis of A-form of monoamine oxidase. Japan J Pharmacol 31: 763—770

Ekstedt B, Oreland L (1976) Effect of lipid depletion on the different forms of monoamine oxidase in rat liver mitochondria. Biochem Pharmacol 25: 119—124

Erwin VG, Dietrich RA (1971) The labeling in vivo of monoamine oxidase by ^{14}C-pargyline: A tool for studying the synthesis of the enzyme. Mol Pharmacol 7: 219—228

Erwin VG, Hellerman L (1967) Mitochondrial monoamine oxidase. I. Purification and characterization of the bovine kidney enzyme. J Biol Chem 242: 4230—4238

Fowler JW, Ross SB (1984): Selective inhibitors of monoamine oxidase A and B: biochemical, pharmacological, and clinical properties. Medicinal Res Revs 4: 323—358

Gomes B, Naguwa G, Kloepfer HG, Yasunobu KT (1969) Amine oxidase. XIV. Isolation and characterization of the multiple beef liver components. Arch Biochem Biophys 132: 16—27

Goridis C, Neff NH (1971) Monoamine oxidase: an approximation of turnover rates. J Neurochem 18: 1673—1682

Hellerman L, Erwin VG (1968) Mitochondrial monoamine oxidase II: action of various inhibitors for the bovine enzyme. J Biol Chem 243: 5234—5243

Houslay MD, Tipton KF (1973) The nature of the electrophoretically separable forms of rat liver monoamine oxidase. Biochem J 135: 173—186

Huang RH, Faulkner R (1981) The role of phospholipids in the multiple forms of brain monoamine oxidase. J Biol Chem 256: 9211—9215

Husain M, Edmondson DE, Singer TP (1981) Catalytic mechanism of MAO from liver. In: Usdin E, Weiner N, Youdim MBH (eds) Function and regulation of monoamine enzymes. Macmillan, London, pp 477—487

Kenney WC, Nagy J, Salach JI, Singer TP (1979) Structure of the covalent phenylhydrazine adduct of monoamine oxidase. In: Singer TP, Von Korff RW, Murphy DL (eds) Monoamine oxidase: structure, function, and altered functions. Academic Press, New York, pp 25—37

Kochersperger LM, Waguespack A, Patterson JC, Hsieh C-CW, Weyler W, Salach JL, Denney RM (1985) Immunological uniqueness of human monoamine oxidases A and B: new evidence from studies with monoclonal antibodies to human monoamine oxidase A. J Neurosci 5: 2874—2881

Knoll J (1981) Can the suicide inactivation of MAO by deprenyl explain its pharmacological effects? In: Singer TP, Ondarza RN (eds) Molecular basis of drug action. Elsevier/North-Holland, New York, pp 185—201

Krantz A, Kokel B, Sachdeva YP, Salach JI, Detmer K, Claessow A, Sahlberg C (1979) Inactivation of mitochondrial monoamine oxidase by β, γ, δ-allenic amines. In: Singer TP, Von Korff RW, Murphy DL (eds) Monoamine oxidase: structure, function, and altered functions. Academic Press, New York, pp 51—70

Levitt P, Pintar JE, Breakefield XO (1979) Immunocytochemical demonstration of monoamine oxidase B in brain astrocytes and serotonergic neurons. Proc Nat Acad Sci US 79: 6385—6389

Lobenstein Verbeek CL, Jungejan JA, Frank J, Duine JA (1984) Bovine serum amine oxidase: a mammalian enzyme having covalently bound PQQ as prosthetic group. FEBS Letters 170: 305—309

McCauley R, Racker E (1973) Separation of two monoamine oxidases from bovine brain. Mol Cell Biochem 1: 73—81

Maycock AL, Abeles RH, Salach JI, Singer TP (1976) The structure of the covalent adduct formed by the interaction of 3-dimethylamino-1-propyne and the flavin of mitochondrial amine oxidase. Biochemistry 15: 114—125

Minamiura N, Yasunobu KT (1978) Bovine liver monoamine oxidase. A modified purification procedure and preliminary evidence for two subunits and one FAD. Arch Biochem Biophys 189: 418—489

Nagy J, Salach JI (1981) Identity of the active site flavin peptide fragments from the human "A" form and the bovine "B" form of monoamine oxidase. Arch Biochem Biophys 208: 388—394

Naoi M, Yagi K (1980) Effect of phospholipids on beef heart mitochondrial monoamine oxidase. Arch Biochem Biophys 205: 18—26

Navarro-Welch C, McCauley RB (1982) An evaluation of phospholipids as regulators of monoamine oxidase A and monoamine oxidase B activities. J Biol Chem 257: 13645—13649

Oreland L (1971) Purification of pig liver mitochondrial monoamine oxidase. Arch Biochem Biophys 146: 410—421

Paech C, Salach JI, Singer TP (1979) Suicide inactivation of monoamine oxidase by trans-phenylcyclopropylamine. In: Singer TP, Von Korff RW, Murphy DL (eds) Monoamine oxidase: structure, function, and altered functions. Academic Press, New York, pp 39–50

Paech C, Salach JL, Singer TP (1980) Suicide inactivation of monoamine oxidase by trans-phenylcyclopropylamine. J Biol Chem 255: 2700–2704

Paech C, Singer TP (1983) Unpublished data

Pearce LB, Ruth JA (1985) Human brain monoamine oxidase type B: mechanism of deamination as probed by steady-state methods. Biochemistry 24: 1821–1826

Pintar JE, Levitt P, Salach JI, Weyler W, Breakefield XO (1983) Specificity of antisera prepared against pure bovine MAO-B. Brain Res 276: 127–139

Salach JI (1979) Monoamine oxidase from beef liver mitochondria: simplified isolation procedure, properties, and determination of its cysteinyl flavin content. Arch Biochem Biophys 192: 128–137

Salach JI, Singer TP, Castagnoli N Jr, Trevor A (1984) Oxidation of the neurotoxic amine MPTP by monoamine oxidase A and B and suicide inactivation of the enzymes by MPTP. Biochem Biophys Res Commun 125: 831–835

Silverman RB, Hoffman SJ (1979) Mechanism of inactivation of monoamine oxidase by N-cyclopropyl-N-arylalkylamines. In: Singer TP, Von Korff RW, Murphy DL (eds) Monoamine oxidase: structure, function, and altered functions. Academic Press, New York, pp 71–79

Silverman RB, Hoffman SJ, Catus III WB (1980) A mechanism for mitochondrial monoamine oxidase-catalyzed amine oxidation. J Am Chem Soc 102: 7126–7128

Silverman RB (1983) Mechanism of inactivation of monoamine oxidase by trans-2-phenylcyclopropylamine and the structure of the enzyme-inactivator adduct. J Biol Chem 258: 14766–14769

Silverman RB, Zieske PA (1985 a) Mechanism of inactivation of monoamine oxidase by 1-phenylcyclopropylamine. Biochemistry 24: 2128–2138

Silverman RB, Zieske PA (1985 b) 1-Benzylcyclopropylamine and 1-(phenylcyclopropyl)methylamine: an inactivator and a substrate for monoamine oxidase. J Med Chem 28: 1953–1957

Silverman RB, Zieske PA (1986) Identification of the amino acid bound to the labile adduct formed during inactivation of monoamine oxidase by 1-phenylcyclopropylamine. Biochem Biophys Res Commun 135: 154–159

Singer TP (1979) Active site-directed irreversible inhibitors of monoamine oxidase. In: Singer TP, Von Korff RW, Murphy DL (eds) Monoamine oxidase: structure, function, and altered functions. Academic Press, New York, pp 7–24

Singer TP (1985 a) Mitochondrial monoamine oxidase: isolation, assay, properties, mechanism, and unresolved questions. In: Zakim D, Vessey DA (eds) Biochemical pharmacology and toxicology, vol 1. J Wiley, New York, pp 229–263

Singer TP (1985 b) Inhibitors of FAD-containing monoamine oxidases. In: Mondovi B (ed) Structure and function of monoamine oxidases. CRC Press, Boca Raton, pp 220–229

Singer TP (1985 c) Demise and resurrection of misconceptions about monoamine oxidase. In: Kelemen K, Magyar K, Vizi S (eds) Neuropharmacology '85. Akademiai Kiado, Budapest, pp 29–36

Singer TP, Barron ESG (1945) Sulfhydryl enzymes in fat and protein metabolism. J Biol Chem 157: 241–253

Singer TP, Husain M (1982) Monoamine oxidase and its reaction with suicide substrates. In: Massey V, Williams CH (eds) Flavins and flavoproteins. Elsevier/North-Holland, New York, pp 389–401

Singer TP, Salach JI (1981) Interaction of suicide inhibitors with the active site of monoamine oxidase. In: Youdim MBH, Paykel ES (eds) Monoamine oxidase inhibitors: the state of the art. J Wiley, New York, pp 17–29

Singer TP, Salach JI, Crabtree D (1985) Reversible inhibition and mechanism-based irreversible inactivation of monoamine oxidases by MPTP. Biochem Biophys Res Commun 127: 707–712

Singer TP, Salach JI, Castagnoli N Jr, Trevor A (1986) Interactions of the neurotoxic amine 1-methyl-4-phenyl-1, 2, 3, 6-tetrahydropyridine with monoamine oxidases. Biochem J 235: 785–789

Smith D, Filipowicz C, McCauley R (1985) Monoamine oxidase A and monoamine oxidase B activities are catalyzed by different proteins. Biochem Biophys Acta 831: 1–7

Von Korff RW (1979) Monoamine oxidase: unanswered questions. In: Singer TP, Von Korff RW, Murphy DL (eds) Monoamine oxidase: structure, function, and altered functions. Academic Press, New York, pp 1–6

Weyler W, Salach JI (1981) Iron content and spectral properties of highly purified bovine liver monoamine oxidase. Arch Biochem Biophys 212: 147–153

Weyler W, Salach JI (1985) Purification and properties of mitochondrial monoamine oxidase type A from human placenta. J Biol Chem 260: 13199–13207

Westlund KN, Denney RM, Kochersperger LM, Rose RM, Abell CW (1985) Science 230: 181–183

White RL, Smith RA, Krantz A (1983) Differential inactivation of mitochondrial monoamine oxidase by allenic amines. Biochem Pharmacol 32: 3661–3664

Yasunobu KT, Watanabe K, Zeidan H (1979) Monoamine oxidase: some new findings. In: Singer TP, Von Korff RW, Murphy DL (eds) Monoamine oxidase: structure, function, and altered function. Academic Press, New York, pp 251–263

Author's address: Dr. T. P. Singer, Molecular Biology Division, Veterans Administration Medical Center, San Francisco, CA 94121, U.S.A.

Singer, H. (1985) Inhibition of PAF-containing monophilic release, in Leukotrienes and Prostaglandins and Inhibition of lipoxygenase (ed. CRC) ... ...

Singer, H. (1985) Isolation and identification of one component ... Prostaglandins (eds. In Robinson, K. Mähler, K. Vies) and Hermann, ... Rolle, Hormonal, pp. 27–31.

Singer, P.P. Brune, M.C. (1985) Sulfido-cysteines in ... antagonism, J. Biol. Chem. 159, 281–297.

Singer, J.A. Lusin, M.H. et al. ... asthma, indias and ... ... and others, in Mähler, V. Williams, C.J. (ed.) Physiology and Physiology of asthma, Plenum (eds.), New York, pp. 246–261.

Singer, M. Zabel, P. (1981) Interaction of ... inhibition with the PAF of ... monocyte surface, in Smoking, ASM ..., the PAF Mediators, ...

J Neural Transm (1987) [Suppl] 23: 25—35

The catalytic behaviour of monoamine oxidase

K. F. Tipton, Anne-Marie O'Carroll*, and J. M. McCrodden

Department of Biochemistry, Trinity College, Dublin, Ireland

Summary

Evidence concerning the kinetic mechanism of the reaction catalyzed by monoamine oxidase is reviewed with particular reference to the possibility that the double-displacement mechanism followed by other substrates is not operative with benzylamine. The requirement for only one of the two products of the first half-reaction to be released in a double-displacement mechanism indicates that the available evidence does not exclude such a mechanism with benzylamine as the substrate.

Cases in which substrates also act as time-dependent inhibitors are considered. The mechanism that can describe the inhibition and product formation is similar for the compounds MD 780236 and MPTP whereas that describing the effects of high concentrations of 2-phenethylamine is best described by a scheme involving inhibition occuring via an abortive complex.

Introduction

Monoamine oxidase (amine: oxygen oxidoreductase [deaminating] [flavin-containing], EC 1.4.3.4: MAO) catalyses the oxidative deamination of primary amines according to the overall equation:

$$RCH_2 NH_2 + O_2 + H_2O \rightarrow RCHO + NH_3 + H_2O_2 \tag{1}$$

Secondary and tertiary amines in which the amine substituting groups are methyl groups are also substrates (see Blaschko, 1952; Tipton, 1975, for reviews).

* Present address: Laboratory of Clinical Sciences, NIMH, Bethesda, MD 20205, U.S.A.

An understanding of the kinetic behaviour of the enzyme is important not only for helping to understand its reaction mechanism and interactions with inhibitors but also in terms of its behaviour within the cell under physiological conditions. This account will review the evidence concerning the kinetic mechanism of the enzyme in the light of these functions.

Initial rate behaviour

Determinations of the initial rates of the reaction catalyzed by monoamine oxidase preparations from several sources and with a number of different substrates have shown double-reciprocal plots of initial velocity against amine or oxygen concentration at a series of fixed concentrations of the other substrate to give rise to families of apparently parallel lines (Tipton, 1968; Fischer et al., 1968; Oi et al., 1970; Houslay and Tipton, 1973; Fowler and Oreland, 1980; Roth, 1979). It has been pointed out that a feature of such behaviour, regardless of the kinetic mechanism that gives rise to it, is that at low concentrations of the amine substrate the initial velocity will be quite insensitive to fluctuations of the physiological oxygen concentrations (Tipton, 1972). Although the K_m values of the enzyme towards oxygen have been shown to depend on the amine substrate, they are relatively high being close to the oxygen concentration in air-saturated water at 37 °C. This suggests that the enzyme would be working below it's maximum velocity in vivo. It has been argued that a high K_m value is an important feature of the evolution of high enzyme catalytic power (Fersht, 1977) and the possible selective advantage of mechanisms leading to such insensitivity to fluctuations in substrate concentrations has been discussed (Tipton, 1980).

The initial velocity behaviour would be consistent with the enzyme following a double-displacement, or ping-pong, mechanism but further experimental data would be required to establish whether that was indeed the case. Product inhibition studies with the enzyme from pig brain (Tipton, 1968), ox thyroid (Fischer et al., 1968), ox liver (Oi et al., 1970) and rat liver (Houslay and Tipton, 1973, 1975) were consistent with such a pathway being followed although the mechanisms proposed differed in detail, which may reflect differences in the enzyme source, substrate used or assay conditions. In its simplest form such a kinetic mechanism may be described by two half-reactions. In the first of these oxidation of the amine leads to the formation of the aldehyde, ammonia and a reduced form of the enzyme (EH_2):

$$E + Amine + H_2O \longrightarrow EH_2 + Aldehyde + Ammonia \quad (2)$$

The reduced enzyme is then re-oxidized in a second step which generates hydrogen peroxide:

$$EH_2 + O_2 \longrightarrow E + H_2O_2 \quad (3)$$

The demonstration of the first half-reaction (2) in the absence of oxygen (Tipton, 1968; Oi et al., 1970) is consistent with this mechanism and the reduction of the flavin component by the amine substrate has also been demonstrated spectrophotometrically (see Husain et al., 1982).

The kinetic equation that describes the mechanism shown in equations (2) and (3) can be written as:

$$v = \cfrac{V}{1 + \cfrac{K_a}{[Amine]} + \cfrac{K_o}{[Oxygen]}} \quad (4)$$

Where V is the maximum velocity, K_a and K_o are the concentrations of amine and oxygen, respectively, which will give half-maximum velocity at saturating concentrations of the other substrate, and the square brackets indicate concentrations.

The first partial reaction is believed to proceed via an imine intermediate which then reacts with water to yield the amine and aldehyde products:

$$RCH_2NH_2 \longrightarrow RCH : NH + (2\,H) \quad (5)$$
$$RCH : NH + H_2O \longrightarrow RCHO + NH_3 \quad (6)$$

Formation of an imine intermediate would be consistent with the mechanisms proposed for the action of several mechanism-based inhibitors (see Singer, 1979; Singer and Salach, 1981, for reviews) and also with the formation of an imine analogue as the product of derived from the oxidation of 1-methyl-4-phenyl-1, 2, 3, 6-tetrahydropyridine (MPTP) by monoamine oxidase B (See Chiba et al., 1984).

It is not established whether, in all cases, the hydrolysis of the imine occurs before or after it's release from the enzyme surface. However, in the case of soluble preparations of the enzyme from rat liver with benzylamine as the substrate the former must be the case since the aldehyde product is released from the complex with the reduced enzyme but the ammonia remains bound until after oxygen has bound to it (Houslay and Tipton, 1975).

It should be noted that since two products are formed in the first partial reaction, it is only necessary for one of these to be released before the binding of oxygen for a double-displacement mechanism to be obeyed. Such a mechanism simply requires a product release step, or an irreversible step, to occur before the second substrate is bound. It appears that the order of product release is dependent on the environment of the enzyme since with membrane-bound preparations of rat liver monoamine oxidase the kinetic behaviour suggests that ammonia leaves the reduced enzyme before the aldehyde and both products leave the enzyme before oxygen is bound (Houslay and Tipton, 1973).

The kinetic mechanism followed by an enzyme depends on the relative values of individual rate constants for the overall pathway, and thus it is not surpising that altering the conditions may result in changes in the reaction mechanism such as that discussed above. Changes of reaction mechanism have been shown to occur with several other enzymes if the assay conditions are changed; indeed the details of the double-displacement mechanism obeyed by MAO are sensitive to pH (Oi et al., 1971) and a ternary complex mechanism has been reported to be followed during oxidation of the inhibitor 2-phenethylhydrazine (Tipton and Spires, 1971).

Stopped-flow experiments have been used to measure the rates of the two half-reactions separately using highly purified preparations of the enzyme from ox liver (Husain et al., 1982). With 2-phenethylamine as substrate data were entirely consistent with a double-displacement mechanism being followed in which the second half-reaction was rate-determining. The results obtained with benzylamine were less clear-cut in that the rate of reduction of the enzyme-bound flavin was found to be consistent with the rate of the overall reaction whereas the rate of reoxidation of the chemically-reduced enzyme was much slower. These data are not consistent with the reaction of the free reduced enzyme being a part of the catalytic process and thus it was suggested that with this substrate oxygen must interact with the enzyme before the release of products. These studies however did not show whether both products, aldehyde and ammonia, or only one of them remained bound in the second half-reaction and thus as described above, they are not inconsistent with a double-displacement mechanism being followed.

Pearce and Roth (1985) have studied the inhibition of preparations of human brain monoamine oxidase-B by amphetamine they found this compound to be a mixed-type inhibitor of the activities towards 2-phenethylamine, tyramine and tryptamine which would be consistent with amphetamine being able to bind both to the oxidized and reduced forms of the enzyme in a double-displacement mechanism. However amphetamine was found to be a competitive inhibitor with respect to benzylamine and an uncompetitive inhibitor with respect to oxygen when benzylamine was used as the substrate. These results indicate that amphetamine is only able to bind to the free oxidized form of the enzyme when benzylamine is the substrate and would be consistent with either or both of the products from the first half-reaction not being released from the enzyme before the binding of oxygen.

These workers also showed the inhibition of the oxidation of 2-phenethylamine, tryptamine and tyramine by benzylamine to be mixed-type, which would be consistent with a double-displacement mechanism being followed with these substrates in which the inhibitor, benzylamine, was able to bind to both the oxidized and reduced forms of the enzyme. In contrast, however, tyramine and tryptamine were found to be competitive inhibitors of the oxidation of benzylamine consistent with these amines being unable to bind to the reduced form of the enzyme. These results would, again, indicate the kinetic mechanism followed with benzylamine to differ from the simple double-displacement mechanism followed by the other substrates, in that complete dissociation of the products of the first half reaction before oxygen binding does not occur with that substrate. They do not, however, establish whether both the aldehyde and ammonia or only one of them remains bound. Further work would be needed to show which was the case.

Unfortunately no complete kinetic studies have been carried out with the A-form of monoamine oxidase. Initial velocity studies with that form of the enzyme yield the same parellel double-reciprocal plots as those given by the B-form of the enzyme (see Fowler and Oreland, 1980; Roth and Eddy, 1980) which would be consistent with a double-displacement mechanism also operating in this case. However further work is necessary to verify this tentative conclusion.

Specificity

At low concentrations of the amine substrate and with the oxygen concentration close to the K_m value equation (4) reduced to:

$$v = V \text{ [Amine]}/K_a, \tag{7}$$

$$= k_{cat} \text{ [Enzyme] [Amine]}/K_a. \tag{8}$$

The constant k_{cat}/K_a thus represents the apparent second-order rate constant for enzyme-amine combination. If the rate-limiting step in the enzyme-catalyzed reaction were binding of amine to the enzyme this would be the rate constant for that reaction, otherwise it would represent a lower limit to the value of that constant (see Fersht, 1977).

The enzyme concentration may be determined by estimating the binding of the radioactively-labelled mechanism-based inhibitor pargyline (Parkinson and Callingham, 1980; Gomez et al., 1986) and the values of k_{cat}/K_m calculated for several substrates are summarized in Table 1. These values are all considerably lower than those that would be expected for a simple diffusion-controlled substrate binding reaction (see e.g. Fersht, 1977) which would be consistent with the combination of enzyme and amine not corresponding to the rate-determining step in the reaction.

Table 1. Kinetic constants for human brain MAO

Substrate	Form	K_m (μM)	$k$cat (S^{-1})	$k$cat/K_m ($S^{-1} \cdot mM^{-1}$)
Adrenaline	A	208 ± 56	4.0 ± 0.6	19
	B	226 ± 16	4.3 ± 0.7	19
Noradrenaline	A	284 ± 17	5.9 ± 1.0	21
	B	238 ± 30	3.0 ± 0.4	13
Dopamine	A	212 ± 33	7.1 ± 1.2	33
	B	229 ± 33	6.5 ± 1.1	28
Tyramine	A	127 ± 18	1.9 ± 0.2	15
	B	107 ± 21	3.2 ± 0.3	30
5-Hydroxytryptamine	A	137 ± 24	2.4 ± 0.2	18
	B	$1,093 \pm 22$	0.06 ± 0.02	0.05
2-Phenethylamine	A	140 ± 22	0.21 ± 0.05	1.5
	B	4 ± 3	2.9 ± 0.2	730

Homogenates of human cerebral cortex were used for kinetic determinations at pH 7.2 and 37 °C. The concentrations of the two enzyme forms were determined by measuring the binding of radioactively-labelled pargyline (Gomez et al., 1986).

This constant determines the extent of competition between different substrates for metabolism by the same enzyme. Thus for the system;

$$E + P2 \xleftarrow{\;S2\;} E \xrightarrow{\;S1\;} E + P1 \tag{9}$$

The flux between the two competing pathways will be given by:

$$\frac{v_{S1}}{v_{S2}} = \frac{(k_{cat}/K_m)_{S1}\,[S1]}{(k_{cat}/K_m)_{S2}\,[S2]}. \tag{10}$$

In the case of monoamine oxidase the situation is complicated by the presence of two forms of the enzyme in many tissues. For two enzymes competing for the same substrate the relative flux between them will be given by:

$$\frac{v_A}{v_B} = \frac{V_A(K_B + [S])}{V_B(K_A + [S])}.$$ (11)

Where the subscripts A and B refer to the two enzyme species. At high substrate concentrations the relative flux will depend solely on the ratio of maximum velocities whereas at very low substrate concentrations the flux ratio will be given by:

$$\frac{v_A}{v_B} = \frac{(k_{cat}[Enzyme]/K_m)_A}{(k_{cat}[Enzyme]/K_m)_B}.$$ (12)

Thus the competition will depend *inter alia* on the relative concentrations of the two enzyme forms. In rat liver (Gomez *et al.*, 1986) and human brain pargyline binding studies have indicated the ammounts of the two forms present to be similar. However this certainly does not appear to be the case in all other species and tissues (see Mantle and Tipton, 1982; Tipton, 1986 for reviews).

Substrates as inhibitors

Several workers have found the enzyme to be inhibited by high concentrations of the amine substrate (see Williams, 1974; Suzuki *et al.*, 1979; Kinemuchi *et al.*, 1980; Pearce and Roth, 1985). However with benzylamine, and perhaps in some other cases, this has been shown to result from contamination of the substrate with the aldehyde product (Houslay and Tipton, 1973). The situation with the substrate 2-phenethylamine is however more complicated. At higher concentrations of 2-phenethylamine there is a time-dependent inhibition of the enzyme, which subsequent studies have shown to be slowly reversible. This process may be described by a kinetic mechanism of the form (Kinemuchi *et al.*, 1982);

$$SE \xleftarrow{\ S\ } E \xrightarrow{\ S\ } ES \longrightarrow E + Product \qquad\qquad (13)$$
$$\downarrow \nearrow {\scriptstyle S}$$
$$S\text{-}E$$

where the rates of formation and breakdown of the species S-E are relatively slow. The result of this behaviour is that the time-course of product formation is non-linear at high concentrations of 2-phenethylamine, whereas it is also non-linear at low concentrations because of substrate depletion. Failure to take account of this non-linearity at higher substrate concentrations can give rise to apparent high-substrate inhibition. However if the true initial rates of the reaction were measured no such inhibition by 2-phenethylamine was detected (Kinemuchi *et al.*, 1982).

The molecular mechanism of this time-dependent inhibition remains to be resolved. However similar behaviour has been observed with long-chain aliphatic amines (Von Korf and Wolfe, 1984) and time-dependent inhibition by histamine, which is not a substrate for the enzyme, has also been reported (Lyles and Schaffer, 1979).

Inhibitors as substrates

Although the distinction between substrate that act as time-dependent inhibitors of monoamine oxidase and inhibitors that also act as substrates is somewhat arbitary, some compounds that have been developed as inhibitors of the enzyme have also been found to be substrates.

The time-course of such reactions will be non-linear as shown in Fig. 1. After cessation of reaction it will not be restarted by the addi-

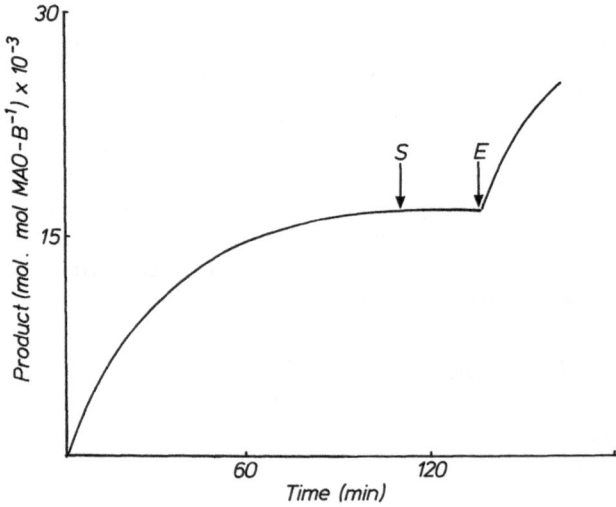

Fig. 1. Time-course of product formation during the enzyme-activated irreversible inhibition of rat liver monoamine oxidase-B by MPTP. Product formation was monitored spectrophotometrically at 37 °C by the procedure described by Fuller and Hemrick-Luecke (1985). At point *S* a further sample of MPTP was added, at point *E* a further sample of enzyme was added

tion of more substrate, or by the addition of a different substrate, but the addition of a further sample of the enzyme will restart the reaction. If inhibition is truely irreversible, dilution or dialysis of the inhibited enzyme will not result in regain of activity. A simple kinetic mechanism that will describe such behaviour is given by the equation:

$$E + I \rightleftharpoons EI \rightleftharpoons (EI)^* \longrightarrow E\text{-}I \qquad (14)$$

$$\downarrow$$

$$E + Products$$

where the E.I, (EI)* and E-I represent the initial, noncovalent, enzyme-inhibitor complex, an activated complex and the irreversibly inhibited complex, respectively.

Since mechanism-based irreversible inhibitors are activated by part of the normal catalytic process of the enzyme (see Singer, 1979; Singer and Salach, 1981) it would not be surprising to find that breakdown of the activated intermediate complex, to yield active enzyme and products, could occur in some cases in addition to reaction within that complex to form an inhibitory adduct.

One such example of this behaviour concerns the compound MD 780236 which acts as both a substrate and an irreversible inhibitor of monoamine oxidase-B with approximately 530 mol of product being formed per mol of enzyme inhibited (Tipton et al., 1983). More recently we have found the neurotoxin MPTP also to obey equation (14), as shown in Fig. 1. Kinetic analysis of the progress curves of the reaction has shown about 17,000 mol of product to be formed per mol of enzyme inhibited in a reaction which has a half-time for inhibition of approximately 21 min.

Acknowledgement

We are grateful to the Medical Research Council of Ireland Unit for Affective Disorders for support.

References

Blaschko H (1952) Amine oxidase and amine metabolism. Pharmacol Rev 4: 415–452
Chiba K, Trevor A, Castagnoli N (1984) Metabolism of the neurotoxic tertiary amine MPTP by brain monoamine oxidase. Biochem Biophys Res Commun 120: 574–578

Fersht AR (1977) Enzyme structure and mechanism. Freeman, Reading, pp 91, 125–133, 244–273

Fischer AG, Schulz AR, Oliner L (1968) Thyroidal biosynthesis of iodo-thyronines II. General characteristics and purification of mitochondrial monoamine oxidase. Biochim Biophys Acta 159: 460–471

Fowler CJ, Oreland L (1980) The nature of the substrate-selective interaction between rat liver monoamine oxidase and oxygen. Biochem Pharmacol 29: 2225–2233

Fuller RW, Hemrick-Luecke SK (1985) Inhibition of types A and B mono-amine oxidase by 1-methyl-4-phenyl-1, 2, 3, 6-tetrahydropyridine. J Pharmacol Exp Ther 232: 696–701

Gomez N, Unzeta M, Tipton KF, Anderson MCJ, O'Caroll A-M (1986) Determination of monoamine oxidase concentrations in rat liver by inhibitor binding. Biochem Pharmacol 35: 4467–4472

Houslay MD, Tipton KF (1973) The reaction pathway of membrane bound rat liver mitochondrial monoamine oxidase. Biochem J 135: 735–750

Houslay MD, Tipton KF (1975) Rat liver mitochondrial monoamine oxidase: A change in reaction mechanism on solubilization. Biochem J 145: 311–321

Husain M, Edmondson DE, Singer TP (1982): Kinetic studies on the catalytic mechanism of liver monoamine oxidase. Biochemistry 21: 595–600

Kinemuchi H, Wakui Y, Kamijo K (1980) Substrate selectivity of type A and B monoamine oxidase in rat brain. J Neurochem 35: 109–115

Kinemuchi H, Arai Y, Oreland L, Tipton KF, Fowler CJ (1982) Time-dependent inhibition of monoamine oxidase by β-phenethylamine. Biochem Pharmacol 31: 959–964

Lyles GA, Shaffer CJ (1979) Substrate specificity and inhibitor sensitivity of monoamine oxidase in rat kidney mitochondria. Biochem Pharmacol 29: 1099–1106

Mantle TJ, Tipton KF (1982) Monoamine oxidase A and B: Time for re-evaluation? In: Kalsner S (ed) Trends in autonomic pharmacology, vol 2. Urban & Schwarzenberg, Baltimore, pp 523–542

Oi S, Shimada K, Inamasu M, Yasunobu KT (1970) Mechanistic studies of beef liver mitochondrial amine oxidase. XVIII. Amine oxidase. Arch Biochem Biophys 145: 28–37

Oi S, Yasunobu KT, Westley J (1971) The effects of pH on the kinetic parameters and mechanism of beef liver monoamine oxidase. Arch Biochem Biophys 145: 557–564

Pearce LB, Roth JA (1985) Human brain monoamine oxidase B: Mechanism of deamination as probed by steady-state methods. Biochemistry 24: 1821–1826

Parkinson D, Callingham BA (1980) The binding of [^{3}H]-pargyline to rat liver mitochondrial monoamine oxidase. J Pharm Pharmacol 32: 49–54

Roth JA (1979) Effect of drugs on inhibition of oxidized and reduced form of MAO. In: Singer TP, Von Korff RW, Murphy DL (eds) Monoamine oxidase: structure function and altered functions. Academic Press, New York, pp 153—168

Roth JA, Eddy BJ (1980) Kinetic properties of membrane-bound and Triton X-100-solubilized human brain monoamine oxidase. Arch Biochem Biophys 205: 260—265

Singer TP (1979) Active-site directed irreversible inhibitors of monoamine oxidase. In: Singer TP, Von Korff RW, Murphy DL (eds) Monoamine oxidase: structure, function and altered functions. Academic Press, New York, pp 7—24

Singer TP, Salach JI (1981) Interaction of suicide inhibitors with the active site of monoamine oxidase. In: Youdim MBH, Paykel ES (eds) Mono-amine oxidase inhibitors—the state of the art. J Wiley, Chichester, pp 17—29

Suzuki O, Katsumata Y, Oya M, Matsumato T (1979) Effect of β-phenyl-ethylamine concentration on its specificity for type A and type B monoamine oxidase. Biochem Pharmacol 28: 953—956

Tipton KF (1968) The reaction pathway of pig brain mitochondrial mono-amine oxidase. Eur J Biochem 5: 316—320

Tipton KF (1972) Some properties of monoamine oxidase. Adv Biochem Psychopharmacol 5: 11—24

Tipton KF (1975) Monoamine oxidase. Handbook Physiol Endocrinol 6: 677—697

Tipton KF (1980) Kinetic mechanism and enzyme function. Biochem Soc Trans 8: 242—245

Tipton KF (1986) Enzymology of monoamine oxidase. Cell Biochem Funct 4: 79—87

Tipton KF, Spires IPC (1971) The kinetics of 2-phenylethylhydrazine oxida-tion by monoamine oxidase. Biochem J 125: 521—524

Tipton KF, Fowler CJ, McCrodden JM, Strolin Benedetti M (1983) The enzyme-activated irreversible inhibition of type-B monoamine oxidase by 3-(4-[(3 chlorophenyl)methoxy]phenyl)-5-[(methylamino) methyl]-2-oxazolidinone methanesulphate (compound MD 780236) and the enzyme-catalyzed oxidation of this compound as competing reactions. Biochem J 209: 235—242

Von Koff RW, Woolfe AR (1984) Saturated amines and diamines as sub-strates which inhibit beef liver mitochondrial monoamine oxidase. J Bioenerg Biomemb 16: 597—609

Williams CH (1974) Monoamine oxidase-I. Specificity of some substrates and inhibitors. Biochem Pharmacol 23: 615—629

Authors' address: Dr. K. F. Tipton, Department of Biochemistry, Trinity College, Dublin 2, Ireland.

J Neural Transm (1987) [Suppl] 23: 37—54

Some properties of semicarbazide-sensitive amine oxidases

B. A. Callingham and M. A. Barrand

Department of Pharmacology, University of Cambridge, Hills Road, Cambridge,
United Kingdom

Summary

The semicarbazide-sensitive amine oxidases (SSAOs) comprise a substantial but diffuse group of enzymes separable from classical monoamine
oxidase in several respects. Differences in cofactor requirement, molecular
weight and subcellular distribution are crucial for such a separation. Differential sensitivity to enzyme inhibitors, characterized by resistance to
inhibition by acetylenic MAO inhibitors coupled with sensitivity to semicarbazide and some related compounds are characteristic of these enzymes.
SSAO enzymes have been found in the plasma of man, ox, pig and horse, for
example as well as in the solid tissues of many species. Extensive studies
have so far failed to produce any conclusive evidence to indicate what the
precise functions of many of these enzymes may be. Indeed in most cases
there is no clear idea as to the nature of the preferred physiological substrate, although many amines with pharmacological activity have been
shown to be substrates. The actions of these amines may be potentiated
following inhibition of SSAO, but as yet little is known whether or not these
actions can be important *in vivo*. An attempt is made in this review to bring
together some of the evidence to see if there are indications for future
endeavours.

Introduction

Soon after the discovery of the enzyme responsible for the oxidative deamination of tyramine in rabbit liver by Hare (1928) another
amine oxidase was found that deaminated histamine (Best, 1929).
From the outset it was apparent that the nature and properties of

these enzymes differed on several counts, which culminated in the crucial decision to call the first enzyme, monoamine oxidase (MAO) and the second, diamine oxidase (DAO, histaminase; Zeller, 1938). This simple division sufficed until suspicions were confirmed that MAO was two enzymes (see Tipton, 1986) and that histaminase was just one of a family of amine oxidases whose common characteristic was a cofactor different from the FAD of MAO (see Blaschko, 1974). Indeed, several of these enzymes have been revealed as the deaminating activity that remains after total inactivation of MAO by acetylenic inhibitors such as clorgyline (Coquil *et al.*, 1973; Lyles and Callingham, 1975). Together with a shared immunity to clorgyline, these enzymes are inhibited by semicarbazide and other so-called carbonyl reagents (see Kapella-Adler, 1970; Blaschko, 1974; Lewinsohn, 1984). However, since MAO and most if not all semicarbazide-sensitive amine oxidases (SSAOs) use oxygen as their second substrate they are classified as EC 1.4.3.4 and EC 1.4.3.6, respectively although the latter classification was originally that of DAO alone (Dixon and Webb, 1967; Banchelli *et al.*, 1983). The SSAOs now comprise an ever-growing group of enzymes, which defy systematic classification and proper naming, a problem fully explored by Lewinsohn (1984) in her comprehensive review concentrating mainly on SSAO in human subjects. This enzyme or enzymes, she argues, should be called benzylamine oxidase since this amine is by far the best substrate but this is not true for all such enzymes in other species. In this review SSAO has been used as a fairly general and imprecise name for this rather nebulous group.

The first clues to the presence of SSAO enzymes appeared when spermine oxidase was discovered in the plasma of sheep and cattle (Hirsch, 1953; Tabor *et al.*, 1954) closely to be followed by benzylamine oxidase in horse serum (Bergeret *et al.*, 1957). Later, human serum and plasma were shown to contain an enzyme now generally referred to as benzylamine oxidase (McEwen and Cohen, 1963; Lewinsohn *et al.*, 1978). In addition to these circulating enzymes can now be added an assortment of tissue SSAOs (see Barrand and Callingham, 1984) and the lysyl oxidases, important in the biosynthesis of collagen and elastin cross-linkage (Blaschko, 1974; Buffoni, 1980; Levene and Carrington, 1985). Distinctions between these various enzymes are not always easy to establish and they may share common substrates. For example, the ox dental pulp enzyme can metabolize both benzylamine and lysyl-vasopressin (Nakano *et al.*, 1974). However, as a rough generalization, SSAO enzymes that are tissue bound or in the circulation usually possess a comparatively high affinity for the monoamine, benzylamine. Such enzymes are not confined solely to the tissues of mammals. They appear to be widely distributed not only through the animal kingdom but also through plants and micro-organisms (see Blaschko, 1974). If tissue histaminase is excluded, it has been the circulating enzymes that have taken historical precedence. However, a great deal of attention is now being lavished on these tissue bound enzymes.

Detection and tissue distribution of SSAOs

It was the discovery of the irreversible MAO inhibitor, clorgyline (Johnston, 1968) that provided a breakthrough in the search for SSAO enzymes by dramatically simplifying the methodology. With crude homogenates of tissues, exposure to millimolar concentrations of clorgyline for around 30 min or so results in the total inhibition of MAO-A and almost all MAO-B (Fowler and Callingham, 1979). Any deaminating activity that then remains is almost certainly due to SSAO. So simple and effective is this method that the residual activity has been called clorgyline-resistant amine oxidase (CRAO) (see Barrand and Callingham, 1982), a name that did not gain approval but for a while had quite a wide currency. It would have been equally justified to call the enzyme deprenyl (selegiline)-resistant amine oxidase since it is largely unaffected by acetylenic inhibitors. It is however sensitive to hydrazine inhibitors but some of these may also inhibit MAO (see Kapella-Adler, 1970; Lyles, 1984). The acetylenic inhibitors of MAO can therefore be used to reveal SSAO activity even in the presence of overwhelming amounts of MAO activity as is seen, for example, in the rat liver (Parkinson et al., 1980).

Together with and in addition to the use of clorgyline, careful selection of substrate and concentration may also reveal the presence of tissue SSAO. If the Km of a substrate is much lower for SSAO than for MAO, careful selection of experimental conditions will allow the SSAO to be detected preferentially and characterized. The SSAO in the rat heart is a particularly good example since the Km of benzylamine for SSAO is about $5\,\mu M$, which is many times lower than its Km for MAO-B (Lyles and Callingham, 1975). The use of $1\,\mu M$ benzylamine will reveal the presence of SSAO while concentrations in the range of $0.5-10\,\mu M$ allow reliable kinetic studies to be undertaken. Care must be taken however, since Lewinsohn in her epic studies of the SSAO activities in human subjects showed clearly that the Km was around $30\,\mu M$ or so and quite close to that of MAO-B (Lewinsohn, 1980, 1984).

Distribution through species and organs is wide, ranging through rat cardiovascular tissue (Coquil *et al.*, 1973), pig dental pulp (Norqvist *et al.*, 1982) to horse liver (R. B. Williams, unpublished) for example (see Lewinsohn, 1984). But what of its subcellular distribution? Does it lamely follow in the footsteps of MAO or does it strike an independent line characterized by it appearing where MAO does not? So far the evidence bearing on this aspect is not prolific.

In tissue fractionation studies of the amine oxidase activities of the rat aorta, a tissue especially rich in SSAO activity, compared with MAO, it was found that the SSAO activity towards tyramine as a substrate was distributed like 5′-nucleotidase and (oligomycin-insensitive) Mg^{2+}-ATPase (Wibo *et al.*, 1980). This clearly supports the contention that the SSAO in this tissue resides in the plasmalemma, which is in stark contrast to the almost exclusively mitochondrial distribution of MAO. Our own work on the SSAO in brown adipose tissue (BAT) of the rat confirmed these observations (Barrand and Callingham, 1982, 1984). In addition we found that a substantial proportion of enzyme activity was also associated with a microsomal fraction. The SSAO could be released into the supernatant by treatment with Triton X-100 without significant inhibition of activity or change in apparent Km value for the deamination of benzylamine. However, removal of the lipid associated with the SSAO resulted in a considerable reduction in activity and an increase in Km, suggesting that the enzyme may need its attachment to membrane elements for full activity. The glycoprotein nature of the SSAO was established from affinity chromatography with either immobilized concanavalin A or Lens culinaris lectin. In this respect the BAT enzyme was found to resemble both ox and pig plasma amine oxidases, which also interact with concanavalin A (Yasunobu *et al.*, 1976; Falk *et al.*, 1983), indicating the presence of mannose- and glucose-containing residues on the enzyme. Such treatment of the SSAO from BAT with these lectins did not however, inhibit the activity of the enzyme. At present therefore, it is not possible, by this approach, to be able to say which way the active centre is facing. However, no search for any possible heterogeneity in carbohydrate content in the enzyme of BAT, like that seen in the pig plasma amine oxidase (Falk *et al.*, 1983), has so far been undertaken. But, since all protein bound carbohydrates in animal cell plasma membranes are located on the external surface (Rothman and Lenard, 1977), it would seem that part of the SSAO must face outwards.

In plasma membranes prepared from isolated adipocytes of BAT a similar enrichment of the activities of SSAO and the outer membrane enzyme, phosphodiesterase I was seen, which lent further support to the location of SSAO in the plasmalemma (Barrand *et al.*, 1984). However, when outer membranes were prepared by binding intact cells to polycation-coated beads, followed by disruption of their contents, the enrichment of phosphodiesterase I activity was much greater than that seen with SSAO. The discrepancy in results from these two preparations may indicate that SSAO is distributed asymmetrically in the plasmalemma. Much more needs to be done to identify the proper nature of such "sidedness" in the position of the enzyme and its active centre. Nonetheless, we are reasonably confident that SSAO is not associated with MAO since their differential distribution can also be shown histochemically (Barrand *et al.*, 1984).

Requirement for cofactors

When the nature of the cofactor of SSAO enzymes is considered, there has been, and still is, considerable uncertainty of its, or their, identity. For many years pyridoxal appears to have been the preferred candidate, but in many cases there has been little supporting evidence. Blaschko in 1974 was very suspicious of such a blanket assumption and inserted pyridoxal between quotation marks. His suspicions have been justified in the discovery that the amine oxidase from bovine serum contains a covalently bound prosthetic group identified as pyrroloquinoline quinone (PQQ; Lobenstein-Verbeek *et al.*, 1984). This raises the possibility that several other SSAOs may contain a similar group.

Another characteristic that ought to demonstrate a commonality between these enzymes is their need for copper. However, while the presence of copper in several enzymes has been demonstrated (Blaschko, 1974), this may not be true for all. Attempts to demonstrate unequivocally the presence of copper in, for example, the SSAO of the BAT of the rat have so far failed (Barrand and Callingham, 1984), although many SSAO enzymes contain copper as an essential constituent (see *e.g.* Barker *et al.*, 1979). Evidence of inhibition of an enzyme by copper chelating agents may be suspect since some, such as cuprizone are carbonyl reagents. Moreover, cyanide, which also chelates copper (see Crabbe *et al.*, 1976), has an inhibitory effect on some but not all SSAO enzymes (see Lyles and Callingham, 1975; Barrand and Callingham, 1982).

Molecular weights of SSAO enzymes and their subunits

Many SSAO enzymes appear to have rather similar molecular weights (see Lewinsohn, 1984), both as the active enzyme and when their subunits are examined. For example, pig plasma amine oxidase has been estimated to have an Mr of 186,000 to 196,000 (Falk, 1983). When exposed to denaturing agents the enzyme then behaves as a single species with an Mr of 95,000 to 97,000 (Barker et al., 1979). Estimation of the Mr of SSAO in BAT of the rat by radiation in-activation analysis in the linear accelerator produced a value of 183,000 ± 5,000 for both solubilized and membrane-bound enzyme, which is quite different from values reported for MAO (Barrand and Callingham, 1982). Solubilized SSAO activity could be eluted from gel filtration columns in the fraction corresponding to an Mr of 160,000 to 180,000, which is in reasonable agreement and similar to that seen in ox and pig plasma and dental pulp enzymes. To determine the subunit Mr of SSAO, use can be made of the powerful irreversible ligand, hydralazine (Lyles et al., 1983). SSAO exposed to radioactively-labelled hydralazine and then subjected to polyacryl-amide gel electrophoresis yields a value of about 94,000 (Barrand and Callingham, 1985), in close agreement with that seen for the pig plasma enzyme.

The fact that the subunit weights are about half the weight of the active enzymes raises the possibility that the subunits are the same. There is some evidence for identical primary structure (Barker et al., 1979). However, it is interesting that the stoichiometry of the inhibi-tion of the pig plasma enzyme by phenylhydrazine, which like hydralazine inactivation of SSAO, interacts at the active site of the enzyme (Lindstrom et al., 1974), shows only 1 mol of hydra-zine per mol of the dimeric enzyme. Does this mean that two identical subunits contain only one active centre? Suggestions have been made, including negative co-operativity, purification inactiva-tion of one but not the other subunit or some flip-flop mechanism for enzyme action, but the mystery remains (Falk, 1983). Alter-natively, it may simply mean that the experimental methods are too insensitive.

Substrate selectivity

When attention is turned to the substrate selectivities of the various SSAO enzymes, no clear picture emerges, with the possible exception that many such enzymes deaminate the physiologically insignificant (as far as we know) amine, benzylamine. In the case of the enzyme in human plasma and smooth muscle, benzylamine is the major and perhaps the exclusive substrate (Lewinsohn *et al.*, 1978; Lewinsohn, 1980, 1981, 1984). This observation raises some intriguing questions about the physiological significance of an enzyme that is in its element deaminating a substrate that is not thought to occur physiologically. It may of course imply that so far its proper substrate has simply not been identified. It is hard to imagine that benzylamine can be important in man in view of its comparatively high Km for the enzyme. In rat and in some other animal species the story is somewhat different. Benzylamine has a low Km but, more importantly, the substrate selectivity is somewhat wider.

There is no doubt that the substrate selectivities of the various SSAO enzymes that have been discovered so far, present a confused and confusing picture. The state of play closely resembles that seen with MAO around the end of the 1960s. There is no clear view whether or not SSAO, even within the same animal, is the same in all tissues. Little is known of the possible effects of changes in the lipid environment or other membrane constituents on the catalytic activity of SSAO. The situation is compounded by the fact that the known substrates of SSAO are also readily deaminated by MAO, often much more efficiently and at higher capacity. It is not always possible to exploit a wide difference in Km values of particular substrates for MAO and SSAO or, with safety, to use selective inhibitors. There is no evidence currently available that identifies the real physiological amine substrate of tissue bound SSAO enzymes. It must however, be borne in mind that SSAO, having a high affinity and low capacity, is unlikely to function as a scavenger of unnecessary or potentially harmful amines, a role well fulfilled by MAO, but may act to supply a product of the deamination process such as the peroxide. If this were the case it might well require a steady and reliable concentration of saturating concentrations of any suitable amine, with control being exercised by some other endogenous agent. In this context it is interesting that, in human subjects, the activity of the plasma benzylamine oxidase may be influenced in this way (Buffoni *et al.*, 1983). It is against this background that the substrate selectivity of SSAO should be examined. Of course it would be exciting to discover an amine with an exquisitely low Km for SSAO. In addition to the possible physiological significance of such a finding would be the more prosaic but equally valuable means of screening for the enzyme without the need for the simultaneous use of MAO inhibitors. It would also greatly reduce the reliance on benzylamine as the common substrate. However, we should not be blind to the possibility that, contrary to current doctrine, benzylamine may be formed physiologically in small amounts that never see the light of day through the efficient action of SSAO. With the current availability of potent irreversible inhibitors of SSAO, the time is probably ripe for a reinvestigation of this whole problem.

The plasma amine oxidases have been resolved largely on the basis of their substrate selectivity. Some examples that are worthy of note include the sheep spermine oxidase (Hirsch, 1953) that deaminates spermine and spermidine but also, and to a much lesser extent, a variety of amines including putrescine, cadaverine, histamine, tyramine and some aliphatic amines such as n-propylamine and n-butylamine. Ruminant spermine oxidase and the horse benzylamine oxidase will also metabolize homocystamine and cystamine (Bergeret and Blaschko, 1957). On the other hand the enzyme of pig plasma is more effective against mescaline than benzylamine and will also deaminate histamine, tyramine and dopamine (Buffoni and Blaschko, 1964, 1971; Buffoni et al., 1977), while in rat and rabbit, the benzylamine oxidase will also metabolize mescaline, dopamine and tyramine but not histamine (McEwen et al., 1966). In human plasma, benzylamine is deaminated at a rate far greater than any other amine (see Lewinsohn, 1984), making the search for its physiological substrate all the more intriguing. However, Lewinsohn points out that the possibility of a substrate being more rapidly deaminated in vivo than in vitro cannot be excluded. It would seem that this aspect is ripe for study with the aid of H.P.L.C. techniques and modern selective inhibitors of MAO and SSAO.

In tissues and organs, the complexity of the picture is limited only by the relatively small number of observations. In our own work we examined a range of amines to see whether or not they could interact with the enzyme in the BAT of the rat by seeing whether or not they could inhibit the deamination of benzylamine (Barrand and Callingham, 1982). Only primary amines caused this effect. Secondary and tertiary amines had virtually no affinity for the enzyme. Although such experiments do not prove that these amines are themselves substrates for SSAO they do suggest that SSAO may be capable of deaminating circulating noradrenaline and dopamine but not adrenaline. More direct experiments with labelled substrates show that in the rat aorta SSAO deaminates tyramine, but in the heart there appears to be no deamination of this amine although in both tissues benzylamine is a ready substrate (Clarke et al., 1982). In the rat, other amines as well as tyramine and benzylamine, are deaminated, including 2-phenethylamine and kynuramine (Coquil et al., 1973; Fuentes and Neff, 1977; Dial and Clarke, 1979; Lewinsohn et al., 1978; Edwards et al., 1979).

It is certainly the case that no easy generalizations are possible concerning substrate selectivity. Fortunately, the easy way out is to try the effect of millimolar clorgyline on the deamination of any chosen substrate by a tissue homogenate. If the inhibition is incomplete, SSAO is probably at work. This is an important control to make, since it cannot be assumed that a particular primary amine is not a substrate. No extrapolation from species to species is possible. For example, 5-hydroxytryptamine is only rarely a substrate for tissue bound SSAO but it is metabolized by some rather exotic tissues such as the giraffe liver (Callingham, 1983). The time is not yet right for an extensive review of the substrate selectivities of tissue bound SSAO enzymes but in a year or so it will be.

Inhibitor selectivity

When one turns to the sensitivity of SSAOs to inhibitors, there appears to be less confusion to some extent due to the fact that the enzymes are classified by their susceptibility to semicarbazide and related inhibitors (see Kapella-Adler, 1970; Blaschko, 1974). Of the hydrazine compounds, phenelzine is very potent but is also an irreversible inhibitor of MAO (Andree and Clarke, 1982; Clarke et al., 1982), benserazide is a tight binding inhibitor (Andree and Clarke, 1982; Lyles and Callingham, 1982) and hydralazine is an agent that, while interacting reversibly with MAO, irreversibly inhibits the SSAO enzymes of the rat (Lyles et al., 1983; Barrand and Callingham, 1985, and see Lyles, 1984).

It has been found that rat aorta SSAO can metabolize allylamine to the toxic aldehyde, acrolein, which may cause the cardiovascular necrosis that follows the administration of allylamine to rats (Nelson and Boor, 1982). This interaction with SSAO has been a stimulus to the synthesis of related compounds that could be potent inhibitors of the enzyme. An example is the compound MDL 72145, which has been shown by Lyles and Fitzpatrick (1985) to be a powerful irreversible inhibitor of the SSAO of rat aorta. Other known MAO inhibitors also interact to inhibit SSAO, for example, amiflamine, its demethylated derivative, FLA 788 (+) and MD 780236 (Kinemuchi et al., 1984). More traditional agents include (+)-amphetamine and the antidysrhythmic agent, mexiletine, which also shows stereoselectivity (Clarke et al., 1982).

Pharmacology

In spite of the ever-growing literature about many aspects of the biology of the SSAO enzymes, there is precious little that deals with the pharmacology, namely the consequences for the organism of enzyme inhibition. Even treatment of individuals with the powerful inhibitor, hydralazine does not lead to effects that can be accounted for by an exclusive action on SSAO. Indeed, when radiolabelled hydralazine is administered to rats, widespread binding is observed that cannot be associated with SSAO (Baker *et al.*, 1985). There is a clear need for a much more selective ligand, which may be provided by an allylamine derivative rather than a hydrazine.

However, these inhibitors may exert actions on tissues *in vitro* that can be revealed by simple isolated organ techniques. A suitable tissue for such a study is the isolated anococcygeus muscle of the rat (Gillespie, 1972). This tissue contracts to a variety of directly- and indirectly-acting sympathomimetic amines, many of which are sub-

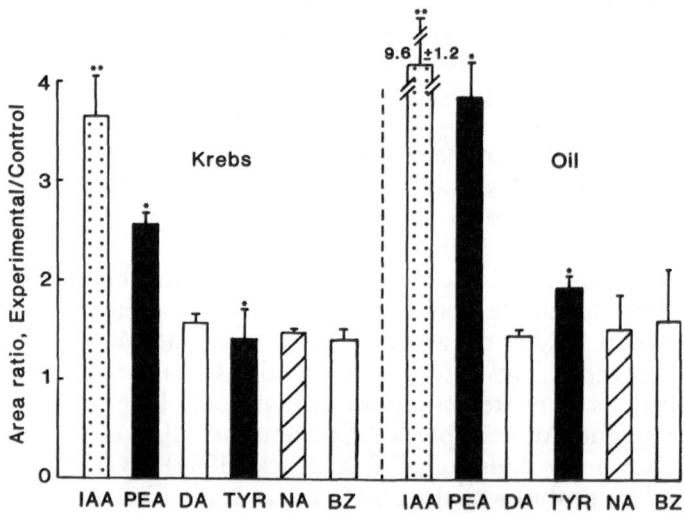

Fig. 1. The effect of hydralazine (5×10^{-6} M) on the isometric tension responses of the isolated anococcygeus muscle of the rat to approximately (ED_{50}) concentrations of a variety of agonists. The histograms represent the areas under the tension responses in the presence of hydralazine, expressed as ratios of the areas of the responses in the absence of the inhibitor. A ratio of 1 would indicate no change in area. Agonists shown are: isoamylamine *(IAA)*, phenethylamine *(PEA)*, dopamine *(DA)*, tyramine *(TYR)*, noradrenaline *(NA)* and benzylamine *(BZ)*. Differences are significant at: * $p < 0.05$, ** $p < 0.01$. Responses measured in: left hand panel; Krebs' solution, right hand panel; oxygenated mineral oil (McCarry, unpublished)

strates for MAO and for SSAO (see Callingham, 1982). Preliminary evidence (Callingham et al., 1984; McCarry, 1986) has shown that inhibition of SSAO by agents such as semicarbazide, benserazide and hydralazine can potentiate the contractile responses to a variety of these amines. Fig. 1 illustrates the effect of 5×10^{-6} M hydralazine on the areas under the responses to amines, both in normal physiological saline and in oxygenated mineral oil. The mineral oil technique of Kalsner and Nickerson (1969) was used to reduce the diffusion of the various agents into the bulk phase of the bathing liquid and integrated responses were felt to be a better measure of any increase in the persistence of an agonist in the biophase. The lack of potentiation of benzylamine by hydralazine is a disappointment, especially since semicarbazide under the same conditions does potentiate (see Callingham, 1986). This may relate to the high concentrations of benzylamine that are needed to cause a response or more likely to other actions of hydralazine.

Physiological role

At the present time there has been no really convincing evidence for an important physiological role of the tissue SSAO enzymes although some suggestions have been made. Even in the case of the plasma benzylamine and spermine oxidases there is still a long way to go before their importance can be established. It is all very well to demonstrate some pharmacology of these enzymes under rather peculiar experimental conditions that may not resemble those pertaining in real life but we need to know if these enzymes impart any benefit to man and other animals. Even MAO-A and B, with all the vast amount of work that has been expended on them have yet to reveal all their mysteries, so it is small wonder that the tissue and plasma SSAO enzymes, rather overwhelmed by their more glamorous cousins are still in the dark ages. But we have some clues.

It is of course well known that many foodstuffs contain potentially harmful amines such as tyramine (see Dostert, 1984) from which the animal is largely protected by the MAO in the intestines and associated structures. However, many animals rely on the absorption of bacterial fermentation products in their normal digestion of nitrogenous materials and cellulose breakdown products, which takes place in the divided stomachs of ruminants and the enlarged caecum of other herbivores. However, these fermentation chambers are also believed to be the source of some dangerous polyamines, such as spermine, a known growth factor for micro-organisms (Kihara and Snell, 1957; Tabor *et al.*, 1961; Tabor and Tabor, 1964). This had led to the suggestion that the spermine oxidase of ruminant plasma is important in their metabolism (Blaschko and Bonney, 1962), especially since there is evidence that, in phylogenesis, spermine oxidase and the rumen may have developed in parallel (Blaschko and Hawes, 1959). The amine oxidases may also help to protect the animal from the vaso-active amines that are formed in response to the lipo-polysaccharides released on the death of pathogenic bacteria (Garner *et al.*, 1978; Moore *et al.*, 1981).

It is hard to decide whether or not, in the human subject, such actions are important, but there are some pointers. They follow from the observations of Lewinsohn (1977, 1984) who demonstrated that the circulating benzylamine oxidase activity was greatly reduced in patients either suffering from severe burns or a variety of malignant disorders. These observations were exciting through the fact that the fall in enzyme activity was closely related to the severity of the disease. In fact the level of the enzyme was, in the case of burning, an important indicator of the subsequent outcome. Moreover the activity recovered after wound healing or after removal of tumour. Changes in enzyme activity have been observed in a variety of patho-logical conditions including chronic liver disease and diabetes (McEwen and Castell, 1967, and see Lewinsohn, 1984) and in experi-mental diabetes in the rat (Hayes and Clarke, 1983). Such associa-tions between enzyme activity and disease are important from the aspect of the chemical pathology but they may also provide a clue to the origin and the real function of this enzyme especially when it appears that it may also be involved in more normal events such as pregnancy (see Lewinsohn, 1984). But are these changes in enzyme activity a reflection of increased protein metabolism elsewhere and what is its relation to the tissue bound enzymes? There are many intriguing questions.

In the case of the tissue bound SSAO enzymes, it seems unlikely that their prime action is one of protection against potentially inju-rious amines. On the whole their capacity and substrate selectivity are limited and there must be better ways available. Since the plasma-lemma of vascular smooth muscle and BAT cells provide rich sources of SSAO, it is tempting to suggest that the enzyme possesses a role in membrane function. It is possible that this may relate in some way to the fact that one of the products of the enzyme reaction, at least in the BAT of the rat, is hydrogen peroxide (Barrand and Callingham, 1984). In white adipose tissue it has been suggested that peroxide may be involved in transmembrane signalling (Mukherjee and Mukherjee, 1982). Indeed, it may be that we have for too long been obsessed with what these enzymes do to amine substrates and have given too little thought to actions of the products other than those directly involved with the enzymology. Previously the peroxide generated by MAO has been shown to be important and not simply a by-product to be mopped up by catalase. For example, the peroxide may be involved in iodothyronine synthesis in the thyroid gland (Fischer *et al.*, 1966), while in the brain it may well influence the synthesis of prostaglandins (Seregi *et al.*, 1982, 1983). The possible involvement of SSAO in membrane signalling may impart a role on the enzyme in some aspect of the maintenance and control of vasomotor tone as well as the responses of adipose tissue to physiological stimuli. Finally, although there is evidence separating tissue SSAO enzymes of the kind described above from lysyl oxidase, there may be some relation between them and even a part for SSAO may be found in collagen and elastin metabolism, especially as the enzyme can be found in rat articular cartilage (Lyles and Bertie, 1986).

After a cursory survey of a highly selected sample of the available literature concerned with SSAO enzymes, one conclusion is obvious. Speculation based on limited evidence can be an enthralling pastime but has precious little merit. It seems to us however, that research into SSAO has at last begun to be recognized as important. Professor Buffoni at the first Amine Oxidase Club meeting in Cambridge in 1984 (Buffoni, 1984) felt that so long after the discovery of spermine oxidase by Hirsch in 1953 times were beginning to change. We must now follow Gaddum's dictum, which freely paraphrased says, "Enough of the talking, now go and do the experiments".

References

Andree TH, Clarke DE (1982) Characteristics and specificity of phenelzine and benserazide as inhibitors of benzylamine oxidase and monoamine oxidase. Biochem Pharmacol 31: 825–830

Baker JRJ, Bullock GR, Williamson IHM (1985) Autoradiographic study of the distribution of [^{3}H]- and [^{14}C]-hydrallazine in the rat. Br J Pharmacol 84: 107–120

Banchelli G, Fantozzi R, Ignesti G, Pirisino R, Raimondi L, Buffoni F (1983) The amine oxidase of human blood lymphocytes and granulocytes. Biochem Int 7: 89–93

Barker R, Boden N, Cayley G, Charlton SC, Henson R, Holmes MC, Kelley ID, Knowles PF (1979) Properties of cupric ions in benzylamine oxidase from pig plasma as studied by magnetic resonance and kinetic methods. Biochem J 177: 289–302

Barrand MA, Callingham BA (1982) Monoamine oxidase activities in brown adipose tissues of the rat: some properties and subcellular distribution. Biochem Pharmac 31: 2177—2184

Barrand MA, Callingham BA (1984) Solubilization and some properties of a semicarbazide-sensitive amine oxidase in brown adipose tissue of the rat. Biochem J 222: 467—475

Barrand MA, Callingham BA (1985) The interaction of hydralazine with a semicarbazide-sensitive amine oxidase in brown adipose tissue of the rat. Biochem J 232: 415—423

Barrand MA, Callingham BA, Fox SA (1984) Amine oxidase activities in brown adipose tissue of the rat: identification of semicarbazide-sensitive (clorgyline-resistant) activity at the fat cell membrane. J Pharm Pharmacol 36: 652—658

Bergeret B, Blaschko H (1957) The oxidation of cystamine and homocystamine by mammalian enzymes. Br J Pharmacol Chemother 12: 513—516

Bergeret B, Blaschko H, Hawes R (1957) Occurrence of an amine oxidase in horse serum. Nature (Lond) 180: 1127—1128

Best CH (1929) The disappearance of histamine from autolysing lung tissue. J Physiol (Lond) 67: 256—263

Blaschko H (1974) The natural history of amine oxidases. Rev Physiol Biochem Pharmacol 70: 83—148

Blaschko H, Bonney R (1962) Spermine oxidase and benzylamine oxidase. Distribution, development and substrate specificity. Proc Roy Soc (Lond) B 156: 268—279

Blaschko H, Hawes R (1959) Observations on spermine oxidase of mammalian plasma. J Physiol (Lond) 145: 124—131

Buffoni F (1980) Some contributions to the problem of amine oxidase. Pharmacol Res Commun 12: 101—114

Buffoni F (1984) Semicarbazide-sensitive amine oxidase: an introductory survey. J Pharm Pharmacol 36: 21 W

Buffoni F, Blaschko H (1964) Benzylamine oxidase and histaminase: purification and crystallization of an enzyme from pig plasma. Proc Roy Soc (Lond) B 161: 153—167

Buffoni F, Blaschko H (1971) Amine oxidase (pig plasma). In: Colowick SP, Kaplan NO (eds) Methods in enzymology, vol 17. Academic Press, New York, pp 682—686

Buffoni F, Della Corte L, Hope DB (1977) Immunofluorescence histochemistry of porcine tissues using antibodies to pig plasma amine oxidase. Proc Roy Soc (Lond) B 195: 417—423

Buffoni F, Banchelli G, Ignesti G, Pirisino R, Raimondi L (1983) The presence of an inhibitor of benzylamine oxidase in human blood plasma. Biochem J 211: 767—769

Callingham BA (1982) Resolution of amine oxidase activities by irreversible inhibitors. In: Kamijo K, Usdin E, Nagatsu T (eds) Monoamine oxidase inhibitors. Basic and clinical frontiers. Excerpta Medica, Amsterdam, pp 100—110

Callingham BA (1983) Comparative aspects of monoamine oxidase. Vet Res Commun 7: 325–330

Callingham BA (1986) Some aspects of monoamine oxidase pharmacology. Cell Biochem Funct 4: 99–108

Callingham BA, McCarry WJ, Barrand MA (1984) Effects of some carbonyl reagents and short-acting MAO inhibitors on semicarbazide-sensitive (clorgyline-resistant) amine oxidase in the rat. In: Tipton KF, Dostert P, Strolin Benedetti M (eds) Monoamine oxidase and disease: prospects for therapy with reversible inhibitors. Academic Press, London, pp 595–596

Clarke DE, Lyles GA, Callingham BA (1982) A comparison of cardiac and vascular clorgyline-resistant amine oxidase and monoamine oxidase. Biochem Pharmacol 31: 27–35

Coquil JF, Goridis C, Mack G, Neff NH (1973) Monoamine oxidase in rat arteries: evidence for different forms and selective localization. Br J Pharmacol 48: 590–599

Crabbe MJC, Waight RD, Bardsley WG, Barker RW, Kelly ID, Knowles PF (1976) Improved purification and characterization of a copper- and manganese-containing amine oxidase with novel substrate specificity. Biochem J 155: 679–687

Dial EJ, Clarke DE (1979) Rat and human cardiac monoamine oxidase: a comparison with other tissues. Eur J Pharmacol 58: 131–319

Dixon M, Webb EC (1967) Enzymes, 2nd edn. Longmans, London

Dostert P (1984) Myth and reality of the classical MAO inhibitors, reasons for seeking a new generation. In: Tipton KF, Dostert P, Strolin Benedetti M (eds) Monoamine oxidase and disease: prospects for therapy with reversible inhibitors. Academic Press, London, pp 9–24

Edwards DJ, Pak KY, Venetti MC (1979) Developmental aspects of rat heart monoamine oxidase. Biochem Pharmacol 28: 2337–2343

Falk MC (1983) Stoichiometry of phenylhydrazine inactivation of pig plasma amine oxidase. Biochemistry 22: 3740–3745

Falk MC, Staton AJ, Williams TJ (1983) Heterogeneity of pig plasma amine oxidase: molecular and catalytic properties of chromatographically isolated forms. Biochemistry 22: 3746–3751

Fischer AG, Schultz AR, Oliner L (1966) The possible role of thyroid monoamine oxidase in iodothyronine synthesis. Life Sci 5: 995–1002

Fowler CJ, Callingham BA (1979) The inhibition of rat heart type A monoamine oxidase by clorgyline as a method for the estimation of enzyme active centers. Mol Pharmacol 16: 546–555

Fuentes JA, Neff NH (1977) Inhibition by pargyline of cardiovascular amine oxidase activity. Biochem Pharmacol 26: 2107–2112

Garner HE, Moore JN, Johnson JH, Clark L, Amend JF, Tritschler LG, Coffman JR, Sprouse RF, Hutcheson DP, Salem CA (1978) Changes in the caecal flora associated with the onset of laminitis. Equine Vet J 10: 249–252

Gillespie JS (1972) The rat anococcygeus muscle and its response to nerve stimulation and to some drugs. Br J Pharmacol 45: 404–416

Hare MLC (1928) Tyramine oxidase I. A new enzyme in liver. Biochem J 22: 968–979

Hayes BE, Clarke DE (1983) Elevation of serum benzylamine oxidase activity in diabetic rats. Proc West Pharmacol Soc 26: 119–122

Hirsch JG (1953) Spermine oxidase: an amine oxidase with specificity for spermine and spermidine. J Exp Med 97: 345–355

Johnston JP (1968) Some observations upon a new inhibitor of monoamine oxidase in brain tissue. Biochem Pharmacol 17: 1285–1297

Kalsner S, Nickerson M (1969) Effects of reserpine on the disposition of sympathomimetic amines in vascular tissue. Br J Pharmacol 35: 394–405

Kapella-Adler R (1970) Amine oxidases and methods for their study. Wiley Interscience, New York

Kihara H, Snell EE (1957) Spermine and related polyamines as growth stimulants for Lactobacillus casei. Proc Nat Acad Sci 43: 867–871

Kinemuchi H, Morikawa F, Ueda T (1984) Inhibition of benzylamine oxidase by some selective monoamine oxidase inhibitors. In: Tipton KF, Dostert P, Strolin Benedetti M (eds) Monoamine oxidase and disease: prospects for therapy with reversible inhibitors. Academic Press, London, pp 574–575

Levene CI, Carrington MJ (1985) The inhibition of protein-lysine 6-oxidase by various lathyrogens. Evidence for two different mechanisms. Biochem J 232: 293–296

Lewinsohn R (1977) Human serum amine oxidase. Enzyme activity in severely burnt patients and in patients with cancer. Clin Chim Acta 81: 247–256

Lewinsohn R (1980) Benzylamine oxidase: an enzyme in search of a function. Ph D thesis, University of London

Lewinsohn R (1981) Amine oxidase in human blood vessels and non-vascular smooth muscle. J Pharm Pharmacol 33: 569–575

Lewinsohn R (1984) Mammalian monoamine-oxidizing enzymes, with special reference to benzylamine oxidase in human tissues. Brazil J Med Biol Res 17: 223–256

Lewinsohn R, Böhm K-H, Glover V, Sandler M (1978) A benzylamine oxidase distinct from monoamine oxidase B—widespread distribution in man and rat. Biochem Pharmacol 27: 1857–1863

Lindstrom A, Olsson B, Pettersson G (1974) Kinetics of the interaction between pig-plasma benzylamine oxidase and hydrazine derivatives. Eur J Biochem 42: 177–182

Lobenstein-Verbeek CL, Jongejan JA, Frank J, Duine JA (1984) Bovine serum amine oxidase: a mammalian enzyme having covalently bound PQQ as prosthetic group. FEBS Letters 170: 305–309

Lyles GA (1984) The interaction of semicarbazide-sensitive amine oxidase with MAO inhibitors. In: Tipton KF, Dostert P, Strolin Benedetti M (eds) Monoamine oxidase and disease: prospects for therapy with reversible inhibitors. Academic Press, London, pp 547–556

Lyles GA, Bertie KH (1987) Properties of a semicarbazide-sensitive amine oxidase in rat articular cartilage. Acta Pharmacol Toxicol (in press)

Lyles GA, Callingham BA (1975) Evidence for a clorgyline-resistant monoamine metabolizing activity in the rat heart. J Pharm Pharmacol 27: 682–691

Lyles GA, Callingham BA (1982) *In vitro* and *in vivo* inhibition by benserazide of clorgyline-resistant amine oxidases in rat cardiovascular tissues. Biochem Pharmacol 31: 1417–1424

Lyles GA, Fitzpatrick CMS (1985) An allylamine derivative (MDL 72145) with potent irreversible inhibitory actions on rat aorta semicarbazide-sensitive amine oxidase. J Pharm Pharmacol 37: 329–335

Lyles GA, Garcia-Rodrigues J, Callingham BA (1983) Inhibitory actions of hydralazine upon monoamine oxidizing enzymes in the rat. Biochem Pharmacol 32: 2515–2521

McCarry WJ (1986) The influence of monoamine oxidase inhibitors on some pharmacological responses of the rat anococcygeus muscle. Ph D thesis, University of Cambridge

McEwen Jr CM, Castell DO (1967) Abnormalities of serum monoamine oxidase in liver disease. J Lab Clin Med 70: 36–47

McEwen Jr CM, Cohen JD (1963) An amine oxidase in normal human serum. J Lab Clin Med 62: 766–776

McEwen Jr CM, Cullen KT, Sober AJ (1966) Rabbit serum monoamine oxidase. I. Purification and characterization. J Biol Chem 241: 4544–4556

Moore JN, Garner HE, Coffman JR (1981) Haematological changes during development of acute laminitis hypertension. Equine Vet J 13: 240–242

Mukherjee SP, Mukherjee C (1982) Similar activities of nerve growth factor and its homologue proinsulin in intracellular hydrogen peroxide production and metabolism in adipocytes. Transmembrane signalling relative to insulin-mimicking cellular effects. Biochem Pharmacol 31: 3163–3172

Nakano G, Harada M, Nagatsu T (1974) Purification and properties of an amine oxidase in bovine dental pulp and its comparison with serum amine oxidase. Biochim Biophys Acta 341: 366–377

Nelson TJ, Boor PJ (1982) Allylamine cardiotoxicity—IV. Metabolism to acrolein by cardiovascular tissues. Biochem Pharmacol 31: 509–514

Norqvist A, Oreland L, Fowler CJ (1982) Some properties of monoamine oxidase and a semicarbazide sensitive amine oxidase capable of the deamination of 5-hydroxytryptamine from porcine dental pulp. Biochem Pharmacol 31: 2739–2741

Parkinson D, Lyles GA, Browne BJ, Callingham BA (1980) Some factors influencing the metabolism of benzylamine by type A and B monoamine oxidase in rat heart and liver. J Pharm Pharmacol 32: 844–850

Rothman JE, Lenard J (1977) Membrane asymmetry. Science 195: 743–753

Seregi A, Serfözö P, Mergl Z (1983) Evidence for the localization of hydrogen peroxide-stimulated cyclooxygenase activity in rat brain mitochondria: a possible coupling with monoamine oxidase. J Neurochem 40: 407–413

Seregi A, Serfözö P, Mergl Z, Schaefer A (1982) On the mechanism of the involvement of monoamine oxidase in catecholamine-stimulated prostaglandin biosynthesis in particulate fraction of rat brain homogenates: role of hydrogen peroxide. J Neurochem 38: 20–27

Tabor H, Tabor CW (1964) Spermidine, spermine and related amines. Pharmac Rev 16: 245–300

Tabor H, Tabor CW, Rosenthal SM (1954) Purification of amine oxidase from beef plasma. J Biol Chem 208: 645–661

Tabor H, Tabor CW, Rosenthal SM (1961) The biochemistry of the poly-amines: spermidine and spermine. Ann Rev Biochem 30: 579–604

Tipton KF (1986) Enzymology of monoamine oxidase. Cell Biochem Funct 4: 79–87

Wibo M, Duong AT, Godfraind T (1980) Subcellular location of semi-carbazide-sensitive amine oxidase in rat aorta. Eur J Biochem 112: 87–94

Yasunobu KT, Ishizaki H, Minamiura N (1976) The molecular, mechanistic and immunological properties of amine oxidases. Mol Cell Biochem 13: 3–19

Zeller EA (1938) Über den enzymatischen Abbau von Histamin und Diaminen. Helv Chim Acta 21: 881–890

Authors' address: Prof. Dr. B. A. Callingham, Department of Pharmacology, University of Cambridge, Hills Road, Cambridge, CB2 2QD, United Kingdom.

J Neural Transm (1987) [Suppl] 23: 55—72

Amine oxidases and their endogenous substrates

(with special reference to monoamine oxidase and the brain)

P. C. Waldmeier

Research Department, Pharmaceuticals Division, Ciba-Geigy Ltd., Basle, Switzerland

Abbreviations

5-HIAA   5-hydroxyindoleacetic acid
5-HT   serotonin
A   adrenaline
BzO   benzylamine oxidase
CG   clorgyline
DA   dopamine
DAO   diamine oxidase
DOPAC   3, 4-dihydroxyphenylacetic acid
DP   deprenil
GABA γ-aminobutyric acid
HVA   homovanillic acid
MAO   monoamine oxidase
MHPG   3-methoxy-4-hydroxyphenylglycol
NA   noradrenaline
PAO   polyamine oxidase
PEA β-phenylethylamine

Summary

The roles of MAO, BzO, DAO and PAO in the metabolism of endogenous substrates and the functional implications of their action and inhibition is reviewed, the emphasis being on MAO on one hand and on brain on the other. The major issues are the following:

1. There is no discrete subdivision into substrates selective for MAO-A, MAO-B, or mixed ones, but rather a continuum.

2. Tissue differences in substrate specificity are not likely to be due to molecular variability of MAO. For the deamination of DA, 5-HT and PEA at least, the relative participation of either MAO form in a given tissue is primarily determined by the relative abundance of the two forms; only at 10^{-5} M and above, substrate concentration begins to matter also.

3. In vivo, compartmentation is of paramount importance: since there seems to be more MAO-A than B inside monoaminergic neurons, DA, 5-HT and NA are predominantly metabolized by MAO-A if metabolism occurs mainly intraneuronally. Conversely, since MAO-B is more abundant extra-neuronally, e.g. in glia cells, the relative participation of this form increases if a significant portion of these amines is deaminated outside monoamin-ergic neurons.

4. In vivo, monoamine deamination is reduced concomitantly with the degree of MAO inhibition, whereas signs of increased transmitter function are only observed if enzyme inhibition is at least 80%. This is likely to be the result of the action of compensatory mechanisms such as feedback inhibi-tion of transmitter release and synthesis.

5. BzO is particularly abundant in vascular tissue, lung and bone. Low levels are found in brain. Endogenous substrates and physiological function are not known. DAO also occurs only in minimal amount in brain, if at all. Its principal substrates are histamine and the polyamines, and the disposal of these amines is probably its main function. Of the PAO's, the type of enzyme found in the rat liver attacks the secondary amino groups and may have a more prominent role in the metabolism of polyamines in the brain than in the periphery. Bovine plasma PAO, which attacks primary amino groups, is only found in the serum of ruminants, but not other species. Its function in the metabolism of polyamines is not known.

This contribution reviews the role of MAO or other amine oxida-ses such as BzO, DAO or PAO in the metabolism of endogenous substrates, and the functional implications of the action of these enzymes and of their inhibition. The emphasis is on MAO on the one hand and on the brain on the other. Particular attention is paid to the question which form of MAO is more important in the metabolism of which substrate under what conditions.

After Johnston's suggestion to subdivide MAO into MAO-A and B subgroups, scores of substrates were checked with respect to the form by which they were deaminated. In 1974, it was thought that there were specific substrates for MAO-A, for MAO-B, and mixed ones (Houslay and Tipton, 1974). Subsequently, substrate specificities were investigated in a variety of tissues of many species, using a wide range of substrate concentrations. It became apparent that the relative participations of the two forms of the enzyme in the deamination of many substrates depended on substrate concentration and also differed from one tissue and one species to the other (see *e.g.* Suzuki *et al.*, 1982). At the same time, kinetic studies were made with various substrates in homogenates, in which either MAO-A or MAO-B were selectively blocked by preincubation with low concentrations of CG or DP, and K_m and V_{max} values were determined with respect to the individual forms (Fowler and Tipton, 1984; Suzuki and Matsumoto, 1985). Plots of the ratios V_{max}^A/V_{max}^B against K_m^B/K_m^A for a number of substrates (Fowler and Tipton, 1984) reasonably fall onto a straight line, suggesting that compounds with high ratios are more or less specific for MAO-A and those with a low ratio for MAO-B, and that there is no discrete subdivision into specific A, specific B, and mixed substrates, but rather a continuum.

A calculation of the percentages of 5-HT, DA and PEA deaminated by mixtures of different proportions of MAO-A or B at various substrate concentrations, using V_{max} and K_m values for these amines with respect to the two forms of MAO in rat liver (Tipton *et al.*, 1982; Fowler and Tipton, 1984) show that, for 5-HT and DA, only the proportion of MAO-A and B active centers determines the relative participation of MAO-B in their deamination up to substrate concentrations of about 10^{-5} M. Above that, substrate concentration begins to matter also. For the participation of MAO-A in PEA deamination, the corresponding concentration limit is perhaps 2–3 times lower.

Strolin-Benedetti *et al.* (1983) measured the K_m values of MAO-A and B towards tyramine in several tissues of the rat, by inhibiting one or the other form specifically with DP or CG. They found them to be very similar, and suggested that tissue differences in substrate specificities are more related to different A/B ratios than to molecular variability. Apparent K_m values towards 5-HT and PEA from different tissues which do not possess unfavorable A/B ratios (such as *e.g.* 90% MAO-B when the K_m towards 5-HT is to be determined) indeed agree remarkably, supporting this contention (Table 1). One might speculate that also the turnover numbers per enzyme molecule of either form of MAO for individual substrates are rather stable across tissues and species. If so, the A/B ratio and under certain conditions also substrate concentrations determine the relative participation of either form of MAO in the deamination of any substrate, irrespective of tissue or species (Fowler and Tipton, 1984; Fowler and Ross, 1984).

Table 1. K_m values towards 5-HT and PEA in different tissues of different species

Tissue	K_m (μM)	Reference
5-HT		
Rat liver	173 ± 82	Tipton et al., 1982
Rat cortex	99 ± 14	Garrick and Murphy, 1982
Rat heart	110 ± 10	Yu, 1979
Rat lung	255	Kung and Wilson, 1979
Human cortex	111 ± 9	Garrick and Murphy, 1982
Human placenta	162 ± 82	Garrick and Murphy, 1982
Bovine liver	400 ± 58	Yu, 1979
Bovine brain	93	Achee et al., 1974
Bovine heart	160	Mantle et al., 1976
Bovine adrenal medulla	65	Tipton et al., 1972
Chick liver	263	Suzuki et al., 1980
Chick brain	345	Suzuki et al., 1980
Chick heart	161	Suzuki et al., 1980
Chick kidney	333	Suzuki et al., 1980
Rabbit brain	130	Achee et al., 1974
PEA		
Rat liver	20 ± 3	Tipton et al., 1982
Rat brain	18 ± 2	Kinemuchi et al., 1980
Rat lung	31	Kung and Wilson, 1979
Human brain	18	Roth and Eddy, 1980
Human platelet	15 ± 8	Koide et al., 1981
Bovine liver	60 ± 10	Yu, 1979
Bovine heart	26	Mantle et al., 1976
Chick liver	31	Suzuki et al., 1980
Chick brain	65	Suzuki et al., 1980
Chick heart	56	Suzuki et al., 1980
Chick kidney	49	Suzuki et al., 1980

Estimations of amine concentrations in compartments pertinent to deamination can be made, e.g. for NA (Dahlström et al., 1966; von Euler, 1972; Stjärne, 1975). The diameters of a varicosity, a small dense core synaptic vesicle and a synaptic cleft have been estimated to be about 1 μm, 50 and 20 nm, and their respective volumes about 5×10^{-16}, 6.5×10^{-20} and 3×10^{-17} liters. To stimulate a low affinity synaptic receptor at a concentration of about 10^{-5} M, 250 molecules or 1.5% of the content of one synaptic vesicle would have to be released per impulse. If 90% of this is transported back into the varicosity, it reaches a concentration of about 5×10^{-7} M there. Cerebral NA neurons firing with a frequency of up to 5 Hz, intraneuronal

MAO must degrade about 1100 molecules per second at the above concentration to prevent a build up in the cytoplasm. Calculations based on kinetic parameters and average number of mitochondria per varicosity are probably too much dependent on assumptions to tell whether this is feasible. Alternatively, a concentration of 10^{-5} M is obtained if one assumes that about 1% of the total NA content in a varicosity is in the cytoplasm and that there are about 20 vesicles per varicosity, containing about 15,000 molecules of NA each (in brain, see Koda and Bloom, 1977). The above 5×10^{-7} M and these 10^{-5} M might be upper and lower limits, respectively, and may apply for the other monoamines as well. Since substrate concentration affects the relative participation of the two forms of MAO in their deamination only above 10^{-5} M, and even then not very markedly, one may assume that, under normal conditions, it is not very likely to play an important role in intact systems.

Compartmentation is of paramount importance for the relative participation of MAO-A and B in the deamination of monoamines in intact tissue preparations and *in vivo*. Usually, substrate specificity has been determined by kinetic investigations in homogenates or mitochondrial preparations, tacitly assuming that the distribution of MAO-A and B in a tissue is homogenous. Indeed, it does not vary much in homogenates from different brain areas, but the ratio of MAO-A to MAO-B is higher in synaptosomes than in free mito-chondria (Student and Edwards, 1977), suggesting that there might be more MAO-A inside and more MAO-B outside neurons. More-over, cholinergic nerve endings in rat brain seem to lack MAO (Arnaiz and De Robertis, 1962). As an example, the importance of compartmentation is illustrated for DA, a mixed substrate in homogenates (Yang and Neff, 1974). Rat brain and liver contain about equal amounts of MAO-A and B (Squires, 1972; Fowler and Oreland, 1980). Based on kinetic data from homogenates, about 30% of DA should be deaminated by MAO-B in the brain. A figure of 20% was found in brain homogenates (Waldmeier et al., 1976) and mitochondria (Schoepp and Azzaro, 1981) of rats pretreated with graded doses of CG and DP. However, the deaminated metabolites of DA, HVA and DOPAC, decreased concomitantly with the reduc-tion in the activity of MAO-A and not B in the c. striatum of rats pretreated with these drugs (Waldmeier et al., 1976), and there was no indication that 20% of endogenous DA might be deaminated by MAO-B. Urwyler and von Wartburg (1980), by studying the deamination of 5-HT, DA, PEA and benzylamine in synaptosomes and cell-free mitochondria, demonstrated that the ratio MAO-A/MAO-B was about 3—4 times greater within the synaptosome than outside. Since uptake precedes the inactivation of most of the released DA, and spillover from the vesicles into the cytoplasm is also practically exclusively deaminated inside the neuron, these authors suggested that DA is almost exclusively deaminated by MAO-A in an intact system. Demarest et al. (1980), also found MAO-A inhibition to be the critical event in the deamination of ^{14}C-DA taken up into synaptosomes prepared from the striata of rats pretreated with graded doses of CG and DP and of ^{3}H-DA newly formed from labelled tyrosine after preincubation of synaptosomes with these drugs. Moreover, they found a decrease of MAO-A, but not MAO-B activity in synaptosome-rich preparations of striata after unilateral 6-OHDA lesions and calculated that about 50% of the MAO-A activity in this area resides within dopaminergic neurons, suggesting a preferential intraneuronal location of this form of the enzyme, in agreement with other groups (Student and Edwards, 1977; Urwyler and von Wartburg, 1980; Oreland et al., 1980; Carlsson et al., 1981; Oreland et al., 1983 a, b; Stenström et al., 1985).

In vitro, CG reduced spontaneous as well as K$^+$-evoked formation of deaminated metabolites from ^{3}H-DA newly formed from ^{3}H-tyrosine in rat striatal slices; DP was ineffective, but inhanced the effect of CG. Moreover, it was also effective in the presence of the DA uptake inhibitor nomifensine under depolarizing conditions (Schoepp and Azzaro, 1982). This latter results suggests that DA can indeed be deaminated by MAO-B in post- or extrasynaptic compartments, if the access to dopaminergic neurons is limited. Other authors have also provided evidence for this, using kainic acid or 6-OHDA-induced lesions (see e.g. Schoepp and Azzaro, 1983; Van der Krogt et al., 1983). In particular, glia cells, which contain relatively higher amounts of MAO-B than MAO-A, may be involved.

The almost if not totally exclusive role of MAO-A for DA deamination does not seem to be restricted to the nigrostriatal pathway. CG decreased DOPAC and reduced DA synthesis in striatum, olfactory tubercles, median eminence and posterior pituitary, and decreased plasma prolactin, concomitantly with MAO-A inhibition. DP was only effective in MAO-A inhibiting doses (Demarest and Moore, 1981). Keane et al. (1981), who studied the acute effects on plasma prolactin of a number of MAO-A and B inhibitors at single, selective doses, came to the same conclusions.

Thus, DA, despite being a substrate for both forms of MAO, is exclusively deaminated by MAO-A in the rat brain under normal conditions, because the compartment in which it is deaminated preferentially contains MAO-A. In the human brain, in which MAO-B predominates over MAO-A, DA has been reported to be a substrate only for the A-form (Tipton *et al.*, 1973), only for the B-form (Glover *et al.*, 1977), or for both (Roth and Feor, 1978; Glover *et al.*, 1980; Garrick and Murphy, 1980). Substrate concentration is unlikely to be a determining factor for the relative participation of the two forms, the K_m values of both forms in human brain towards DA being very similar (O'Carroll *et al.*, 1983). On the other hand, compartmentation might play a similar role in human as in the rat brain. Most of the methodologies used in animal studies are not feasible to address this question. Azzaro *et al.* (1985) have tried to circumvent this problem by using the guinea-pig striatum as a model for the human basal ganglia, since in both of these tissues the ratio of MAO-B to MAO-A is $3:1$ (rat: $1:3$). Indeed, pretreatment of guinea pigs with 1 mg/kg s.c. of either CG or DP alone caused similar changes in the levels of DA and its metabolites, and the combination of both was more effective. Similarly, 1 mg/kg s.c. CG or DP caused clearly less inhibition of ^{14}C-DA deamination in the synaptosomes of the pretreated guinea pigs than the combination of both. However, since dose-effect relationships may be different than in the rat, these results could simply mean that the doses of CG and DP were not adequate and should therefore be interpreted with caution.

The enhancement of the effects of L-DOPA in Parkinson's disease by DP seems to confirm that DA is a substrate for the B enzyme in the human brain. DA formed from exogenous DOPA is generated at least partly outside DA neurons and DP is reported to inhibit the reuptake of this amine. Both MAO-A and B occur in post- and extrasynaptic compartments, particularly in glial cells, in which MAO-B is more abundant than MAO-A, and that neuron death such as that occurring in Parkinsonism is followed by glial proliferation. It is therefore plausible that DA formed from exogenous DOPA becomes substantially metabolized by MAO-B particularly in parkinsonian patients, and that, under such circumstances, DP may effectively inhibit this process.

Urinary amine metabolite changes do certainly not accurately reflect alterations of monoamine metabolism in the brain but may still give some overall information. CG (Linnoila *et al.*, 1982) and brofaromine (CGP 11305 A; Waldmeier *et al.*, 1983; Maitre *et al.*, 1984) increased the urinary output of 3-methoxytyramine and reduced that of HVA without affecting phenethylamine excretion, suggesting that at least in some tissues, DA is substantially deaminated by MAO-A.

On the other hand, CG was found to cause a substantially smaller reduction of CSF HVA than pargyline in depressed patients, whereas the effects on MHPG and 5-HIAA were rather similar (Major *et al.*, 1979). Acute CG treatment of rhesus monkeys, considered as a reasonable substitute for humans because of similar MAO-A/B ratios, at a selective dose caused substantial reductions in MHPG, HVA and 5-HIAA in decreasing order. Repeated treatment with a lower dose caused a similar decrease of MHPG, but only minimal changes of 5-HIAA and HVA (Garrick et al., 1984). Analogous experiments with amiflamine (Garrick et al., 1985 a) and cimoxatone (Garrick et al., 1985 b) gave essentially similar results. The differences in the decreases of the 3 monoamine metabolites may be difficult to interpret. The data may indicate that both forms are involved in the deamination of DA in the primate brain.

5-HT, although being deaminated at high concentrations by MAO-B (Fowler and Tipton, 1982), is quite a preferential substrate for MAO-A. This specificity is certainly enhanced by compartmentation in vivo. However, as in the case of DA, this holds only for normal conditions. If 5-HT gets access to post- or extrasynaptic compartments, parts of it may well become deaminated by MAO-B. Thus, inhibition of both MAO-A and B is required to produce behavioural 5-HT effects after 5-HT uptake inhibition (Lassen and Squires, 1977), tryptophan loading (Squires and Lassen, 1974; Green and Youdim, 1975), and reserpine treatment (Wolf et al., 1985). Although potentiation of 5-HTP effects correlates well with MAO-A but not B inhibition, the effect of CG can still be enhanced by DP (Ortmann et al., 1980). Also, nonspecific MAO inhibitors as well as the combination of CG + DP cause more rapid and more marked increases of brain 5-HT than e.g. CG alone (Green and Youdim, 1975).

NA, originally considered to be as preferential a substrate for MAO-A as 5-HT (Goridis and Neff, 1971; Yang and Neff, 1974; Houslay and Tipton, 1974) in rat tissue homogenates, seems to be less specific according to newer results. Particularly in tissues rich in MAO-B, such as primate brain, a substantial contribution to its deamination comes from this form (White and Glassman, 1977; Garrick and Murphy, 1982; O'Carroll, 1984). In vivo, there is not much evidence that MAO-B significantly contributes to NA deamination under normal conditions in the rat brain or in humans, probably due to compartmentation. However, if NA gets access to MAO-B-rich compartments for any reason, it can be deaminated there. Thus, tranylcypromine and the combination of DP and CG cause greater increases in rat brain NA levels than CG alone (Youdim, 1983).

A is a rather selective substrate for the A form in vitro and in vivo (Zeller and Arora, 1979; Fuller and Hemrick-Luecke, 1981; Fuller et al., 1981). The increases in A in several rat brain regions (Fuller and Hemrick-Luecke, 1981; Mefford et al., 1985) and CSF (Rafaelsen et al., 1985) after MAO inhibition were quantitatively much more important than those of the other monoamines, ranging from 200–500%, comparable to those seen with unregulated systems such as trace amines.

It is of interest with respect to the role of MAO and its inhibition for transmitter function that hardly any increase in the brain concentrations of monoamines nor any behavioural or other functional consequences of MAO inhibition can be seen in animals unless MAO is inhibited by 80% or more. It was therefore assumed that MAO occurs in great excess, and that monoamine metabolism remains unaffected at low degrees of inhibition. This does not seem to be quite correct, however. To illustrate this, Fig. 1 as an example shows a plot of rat striatal HVA decreases vs. *ex vivo* MAO-A inhibition in the rat whole brain, obtained with several MAOI at various doses. The result is a straight line passing through the origin, which suggests that the levels of deaminated metabolites reflect MAO-A inhibition in a linear way, whereas neither the levels of the parent

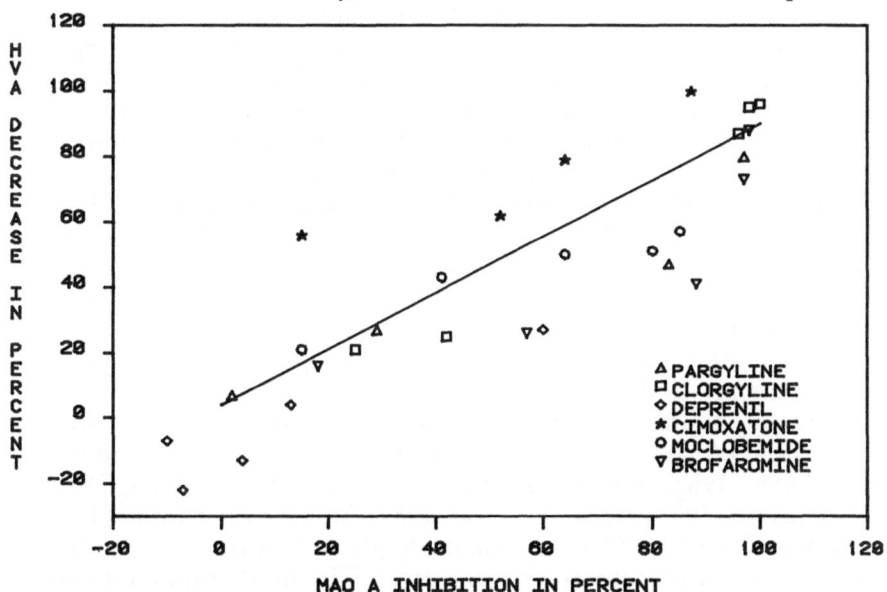

Fig. 1. Correlation between decreases in striatal HVA and MAO-A inhibition in the rat brain after various doses of several MAO inhibitors. HVA levels in the striatum were determined 4 hours after pretreatment of groups of 4–5 rats with various oral or s.c. doses of 6 different MAO inhibitors, by fluorometry or by HPLC with electrochemical detection. MAO-A activity was determined by radioassay after the same pretreatment interval in the rat striatum (pargyline, clorgyline, deprenil) or after 2 hours in the rat whole brain (cimoxatone, moclobemide, brofaromine), using 5-HT as a substrate at a concentration of 20.8 μM (4 animals per group. Linear regression analysis gave a correlation coefficient of 0.839 (n = 28)

amines nor functional activity does so. This may in all probability be ascribed to compensatory mechanisms. Reduction of transmitter synthesis and probably also release gradually compensate for the impairment of monoamine degradation, so that functionally, there is only a net change if the compensatory capabilities of the system are exhausted. This seems to occur only if MAO is inhibited by close to 100%.

BzO (for a recent review see Lewinsohn, 1984), probably the same as clorgyline-resistant amine oxidase, is a pyridoxal-dependent copper-containing amine oxidase, widely distributed in mammalian tissues and particularly abundant in lung, bone and vascular tissue. Very low levels are detected in brain (Lewinsohn et al., 1978; Andree and Clarke, 1981), but brain microvessels are reported to contain very high amounts (Mehrabian and Nalbandyan, 1983). Its best substrate is benzylamine, which does not naturally occur, but other primary amines are also deaminated, although at a much lower rate (McEwen, 1965; Hayes et al., 1983). The endogenous substrates are unknown. It may be of interest that kojic amine, a GABA analogue and GABA agonist, seems to be a substrate for BzO, but not for MAO. Neither GABA nor other GABA agonists interact with the enzyme (Ferkany et al., 1981). Nevertheless, it might be worthwhile to look at some intermediates of the interconversion or metabolism of some amino acids in this respect. The physiological function of BzO is unknown as well. The high concentration in vascular tissue and particularly in brain microvessels might suggest that it is important in preventing certain amine substrates from getting out of the blood stream, in particular into the brain.

DAO or histaminase, another copper enzyme, containing 2 copper atoms per molecule and probably also pyridoxal-dependent, is particularly abundant in placenta, in the gastrointestinal tract, liver and spleen of many species and in the kidney of some mammals (see e.g. Zeller, 1963; Kapeller-Adler, 1970; Shaff and Beaven, 1976; Kusche et al., 1978; Hesterberg et al., 1984). While earlier studies have failed to detect DAO in the brain, Shaff and Beaven (1976) have found a minimal activity in some areas. The most important substrates of DAO are histamine, ultimately transformed into imidazoleacetic acid, and the polyamines. Putrescine is deaminated to the corresponding γ-aminobutyric aldehyde, which cyclizes to delta[1]-pyrroline in the absence and is further oxidized to GABA in the presence of aldehyde dehydrogenase (Seiler et al., 1971). 2-pyrrolidone, the lactam of GABA, is also produced, but it is not known whether it is formed directly by oxidation of γ-aminobutyraldehyde or by cyclization of GABA. The turnover of DAO is more rapid than that of MAO: in the rat intestine, a half-life of about 15 hours was measured (Shaff and Beaven, 1976). According to Sourkes and Missala (1981), DAO is rate-limiting for the oxidative metabolism of putrescine in the rat periphery. In the rat and some other species at least, it also plays a major role in the disposal of peripheral histamine. Since DAO, unlike MAO, can diffuse from the intestinal mucosa into the lumen of the gut, it may also have a role in destroying diamines produced by enteral bacteria. A wide variety of drugs including antimalarials, antitrypanosomals, antihypertensives, neuromuscular blockers, antibiotics etc. inhibit DAO activity (Duch et al., 1984; Sattler et al., 1985), and this is now thought to be responsible for the occurrence of "histamine reactions" sometimes observed in severely ill patients in intensive care units or during polypharmacy.

At least four apparently different PAO's have been described in mammalian species: bovine plasma PAO, rat liver PAO, and two human pregnancy-associated PAO's. Human seminal PAO seems to exhibit similar characteristics as the bovine plasma enzyme. Plants, bacteria and fungi also contain PAO's (for review see Morgan, 1985).

Enzymes similar to bovine plasma PAO have been found in the sera of ruminants, but not other species. It is a glycoprotein consisting of two subunits of M_r 90,000 each and contains two copper atoms per molecule, of which only one seems to be required for full catalytic activity. It is believed to contain covalently bound pyrroloquinoline quinone as the prosthetic group. It oxidizes primary amino groups of polyamines, preferentially those of the aminopropyl moiety, and mono-N-acetylated polyamines, but also some monoamines including benzylamine, and is inhibited by isoniazid and iproniazid, carbonyl binding and sulfhydryl reagents, aminoguanidine and the SAM-decarboxylase inhibitor methylglyoxal-bis(guanylhydrazone). Its function in the metabolism of polyamines is not known.

Rat liver PAO, a single polypeptide of M_r 60,000, contains tightly bound FAD as the cofactor and probably iron. In contrast to the bovine plasma enzyme, it attacks the secondary amino groups of spermine and spermidine, but also preferentially those attached to the aminopropyl moiety. The preferred substrates are the N^1-acetylated derivatives of polyamines. The enzyme is not inhibited by isoniazid, copper chelators, pargyline or the SAM-decarboxylase inhibitor MGBG, but by iron chelators and the flavoprotein inhibitor quinacrin.

An elucidation of the significance of this enzyme for polyamine metabolism in vivo has recently become possible with the development of potent, specific, irreversible enzyme-activated inhibitors (Bolkenius et al., 1985). These compounds were found to inhibit PAO in mouse kidney, liver, spleen and brain irreversibly and completely at i.p. doses above 20 mg/kg. Enzyme activity reappeared with a half-life of about 2—3 days in all these organs. The consequences of this inhibition were decreases of putrescine and, to a lesser extent, spermidine levels, and a strong accumulation of N^1-acetylspermidine and a weaker one of N^1-acetylspermine. N^8-acetylspermidine and N-acetylputrescine were not affected. The accumulation of N^1-acetylspermidine was particularly marked in brain. These data indicate that acetylpolyamines are indeed formed under physiological conditions and strenghthen the evidence for the scheme of polyamine interconversion as originally suggested by Seiler (1981). If the greater accumulation of N^1-acetylspermidine in the brain than in the liver of mice treated with one of these inhibitors was not related to transport phenomena, it might indicated that a significant proportion of polyamines is disposed of by DAO in the periphery.

Polyamines are suspected to play key roles in growth and development of cells, synthesis and metabolism of nucleic acids and proteins, regeneration etc., but in none of these areas anything is firmly established. The availability of reasonably specific drugs to inhibit the catabolism of these endogenous compounds can be expected to greatly facilitate a major progress in the elucidation of their physiological or pathophysiological role and the establishment of their metabolic pathways.

References

Achee FM, Togulga G, Gabay S (1974) Studies of monoamine oxidases: properties of the enzyme in bovine and rabbit brain mitochondria. J Neurochem 22: 651—661

Andree TH, Clarke DE (1981) The isolated perfused rat brain preparation in the study of monoamine oxidase and benzylamine oxidase: lack of selective brain perfusion. Biochem Pharmacol 30: 959—965

Arnaiz GR, DeRobertis CD (1962) Cholinergic and noncholinergic nerve endings in the rat brain. II. Subcellular localization of monoamine oxidase and succinate dehydrogenase. J Neurochem 9: 503—508

Azzaro AJ, King J, Kotzuk J, Schoepp DO, Frost J, Schochet S (1985) Guinea pig striatum as a model of human dopamine deamination: the role of monoamine oxidase isoenzyme ratio, localization and affinity for substrate in synaptic dopamine metabolism. J Neurochem 45: 949—956

Bolkenius FN, Bey P, Seiler N (1985) Specific inhibition of polyamine oxidase in vivo is a method for the elucidation of its physiological role. Biochim Biophys Acta 838: 69—76

Carlsson A, Fowler CJ, Magnusson T, Oreland L, Wiberg A (1981) The activities of monoamine oxidase-A and -B, succinate dehydrogenase and acid phosphatase in the rat brain after hemitransection. Naunyn-Schmiedeberg's Arch Pharmacol 316: 51–55

Dahlström A, Häggendal J, Hökfelt T (1966) The noradrenaline content of the varicosities of sympathetic adrenergic nerve terminals in the rat. Acta Physiol Scand 67: 289–294

Demarest KT, Moore KE (1981) Type A monoamine oxidase catalyzes the intraneuronal deamination of dopamine within nigrostriatal, mesolimbic, tuberoinfundibular and tuberohypophyseal neurons in the rat. J Neural Transm 52: 175–187

Demarest KT, Smith DJ, Azzaro AJ (1980) The presence of the type A form of monoamine oxidase within nigrostriatal dopamine-containing neurons. J Pharm Exp Ther 215: 461–468

Duch DS, Bacchi CJ, Edelstein MP, Nichol CA (1984) Inhibitors of histamine metabolism *in vitro* and *in vivo*. Correlations with antitrypanosomal activity. Biochem Pharmacol 33: 1547–1553

Ferkany JW, Andree TH, Clarke DE, Enna SJ (1981) Neurochemical effects of kojic amine, a gabamimetic, and its interaction with benzylamine oxidase. Neuropharmacology 20: 1177–1182

Fowler CJ, Oreland L (1980) The effect of lipid-depletion on the kinetic properties of rat liver monoamine oxidase-B. J Pharm Pharmacol 32: 681–688

Fowler CJ, Tipton KF (1982) Deamination of 5-hydroxytryptamine by both forms of monoamine oxidase by the rat brain. J Neurochem 38: 733–736

Fowler CJ, Ross SB (1984) Selective inhibitors of monoamine oxidase A and B: biochemical, pharmacological and clinical properties. Med Res Rev 4: 323–358

Fowler CJ, Tipton KF (1984) On the substrate specificities of the two forms of monoamine oxidase. J Pharm Pharmacol 36: 111–115

Fuller RW, Hemrick-Luecke SK (1981) Elevation of epinephrine concentration in rat brain by LY 51641, a selective inhibitor of type A monoamine oxidase. Res Commun Chem Pathol 32: 207–221

Fuller RW, Hemrick-Luecke SK, Perry KW (1981) Influence of harmaline on the ability of pargyline to alter catecholamine metabolism in rats. Biochem Pharmacol 30: 1295–1298

Garrick NA, Murphy DL (1980) Species differences in the deamination of dopamine and other substrates for monoamine oxidase in brain. Psychopharmacology 72: 27–33

Garrick NA, Murphy DL (1982) Monoamine oxidase type A: Differences in selectivity towards l-norepinephrine compared to serotonin. Biochem Pharmacol 31: 4061–4066

Garrick NA, Scheinin M, Chang WH, Linnoila M, Murphy DL (1984) Differential effects of clorgyline on catecholamine and indoleamine metabolites in the cerebrospinal fluid of rhesus monkeys. Biochem Pharmacol 33: 1423–1427

Garrick NA, Seppala T, Linnoila M, Murphy DL (1985 a) The effects of ami-flamine on cerebrospinal fluid amine metabolites in the rhesus monkey. Eur J Pharmacol 110: 1—9

Garrick NA, Seppala T, Linnoila M, Murphy DL (1985 b) Rhesus monkey cerebrospinal fluid amine metabolite changes following treatment with the reversible monoamine oxidase type-A inhibitor cimoxatone. Psychopharmacology 86: 265—269

Glover V, Elsworth JD, Sandler M (1980) Dopamine oxidation and its inhibition by (−)-deprenyl in man. J Neural Transm 16: 163—172

Glover V, Sandler M, Owen F, Riley GJ (1977) Dopamine is monoamine oxidase B substrate in man. Nature 265: 80—81

Goridis C, Neff NH (1971) Monoamine oxidase in sympathetic nerves: a transmitter specific enzyme type. Br J Pharmacol 43: 814—818

Green AR, Youdim MB (1975) Effect of monoamine oxidase inhibition by clorgyline, deprenyl or tranylcypromine on 5-HT concentration in rat brain and hyperactivity following subsequent tryptophan administration. Br J Pharmacol 55: 415—422

Hayes BE, Ostrow PT, Clarke DE (1983) Benzylamine oxidase in normal and atherosclerotic human aortae. Exp Mol Pathol 38: 243—254

Hesterberg R, Sattler J, Lorenz W, Stahlknecht CD, Barth H, Crombach M, Weber D (1984) Histamine content, diamine oxidase activity and histamine methyltransferase activity in human tissues: fact or fictions? Agents Actions 14: 325—334

Houslay MD, Tipton KF (1974) A kinetic evaluation of monoamine oxidase activity in rat liver mitochondrial outer membranes. Biochem J 139: 645—652

Kapeller-Adler R (1970) Amine oxidases and methods for their study. Wiley-Interscience, New York

Keane PE, Menager J, Strolin-Benedetti M (1981) The effect of monoamine oxidase A and B inhibitors on rat serum prolactin. Neuropharmacology 20: 1157—1162

Kinemuchi H, Wakui Y, Kamijo K (1980) Substrate selectivity of type A and type B monoamine oxidase in rat brain. J Neurochem 35: 109—115

Koda LY, Bloom FE (1977) A light and electron microscopic study of noradrenergic terminals in the rat dentate gyrus. Brain Res 120: 327—335

Koide Y, Koide N, Ross S, Saaf J, Wetterberg L (1981) Monoamine oxidase in human platelets. Kinetics and methodological aspects. Biochem Pharmacol 30: 2893—2900

Kung HC, Wilson AG (1979) Characterization of rat pulmonary monoamine oxidase. Life Sci 24: 425—432

Kusche J, Lorenz W, Stahlknecht CD, Friedrich A, Schmidt A, Boo K, Reichert G (1978) Diamine oxidase activity in gastric and duodenal mucosa of man and other mammals with special reference to the pyloric junction. Agents Actions 8: 366—371

Lassen JB, Squires RF (1977) Inhibition of both MAO-A and MAO-B required for the production of hypermotility in mice with the 5-HT uptake inhibitors chlorimipramine and femoxetine. Neuropharmacology 16: 485–488

Lewinsohn R (1984) Mammalian monoamine-oxidizing enzymes, with special reference to benzylamine oxidase in human tissues. Braz J Med Biol Res 17: 223–256

Lewinsohn R, Bohm KH, Glover V, Sandler M (1978) A benzylamine oxidase distinct from monoamine oxidase B-widespread distribution in man and rat. Biochem Pharmacol 27: 1857–1863

Linnoila M, Karoum F, Potter WZ (1982) Effect of low-dose clorgyline on 24-hour monoamine excretion in patients with rapidly cycling bipolar affective disorder. Arch Gen Psychiat 39: 513–516

Maître L, Waldmeier PC, Lauber J, Bieck P (1984) Urinary excretion of catecholamine metabolites, tyramine and phenylethylamine in human volunteers after prolonged treatment with CGP 11305 A and tranylcypromine. In: Tipton KF, Dostert P, Strolin-Benedetti M (eds) Monoamine oxidase and disease. Academic Press, London, pp 117–126

Major LF, Murphy DL, Lipper S, Gordon E (1979) Effects of clorgyline and pargyline on deaminated metabolites of norepinephrine, dopamine and serotonin in human cerebrospinal fluid. J Neurochem 32: 229–231

Mantle TJ, Houslay MD, Garrett NJ, Tipton KF (1976) 5-hydroxytryptamine is a substrate for both species of monoamine oxidase in beef heart mitochondria. J Pharm Pharmacol 28: 667–671

McEwen Jr CM (1965) Human plasma monoamine oxidase. I. Purification and identification. J Biol Chem 240: 2003–2010

Mefford IN, Roth KA, Jurik SM, Collman V, McIntire S (1985) Epinephrine accumulation in rat brain after chronic administration of pargyline and LY 51641–comparison with other brain amines. Brain Res 399: 342–345

Mehrabian ZB, Nalbandyan RM (1983) Benzylamine oxidase from brain microvessels. Febs Lett 164: 89–92

Morgan DM (1985) Polyamine oxidases. Biochem Soc Transact 13: 322–326

O'Carroll AM (1984) The oxidation of noradrenaline by the two forms of human brain monoamine oxidase. In: Tipton KF, Dostert P, Strolin-Benedetti M (eds) Monoamine oxidase and disease. Academic Press, London, pp 593–594

O'Carroll AM, Fowler CJ, Phillips JP, Tobbia I, Tipton KF (1983) The deamination of dopamine by human brain monoamine oxidase. Specificity for the two enzyme forms in seven brain regions. Naunyn-Schmiedeberg's Arch Pharmacol 322: 198–202

Oreland L, Fowler CJ, Carlsson A, Magnusson T (1980) Monoamine oxidase-A and -B activity in the rat brain after hemitransection. Life Sci 26: 139–146

Oreland JL, Arai Y, Stenström A, Fowler CJ (1983 a) Monoamine oxidase activity and localization in the brain and the activity in relation to psychiatric disorders. In: Beckmann H, Riederer P (eds) Monoamine oxidase and its selective inhibitors. Karger, Basel (Mod probl pharmacopsychiat, vol 19, pp 246–254)

Oreland JL, Arai Y, Stenström A (1983 b) The activity of deprenyl (selegiline) on intra- and extraneuronal dopamine oxidation. Acta Neurol
Scand 95: 81—85

Ortmann R, Waldmeier PC, Radecke E, Felner A, Delini-Stula A (1980) The
effects of 5-HT uptake- and MAO-inhibitors on L-5-HTP-induced ex-
citation in rats. Naunyn-Schmiedeberg's Arch Pharmacol 311: 185—192

Rafaelsen OJ, Christensen NJ, Gjerris A (1985) Adrenaline and MAO-inhibi-
tion in CSF and brain. Acta Pharmacol Toxicol 56: 98—104

Roth JA, Eddy BJ (1980) Kinetic properties of membrane-bound and triton
X-100-solubilized human brain monoamine oxidase. Arch Biochem
Biophys 205: 260—266

Roth JA, Feor K (1978) Deamination of dopamine and its 3-O-methylated
derivative by human brain monoamine oxidase. Biochem Pharmacol
27: 1606—1608

Sattler J, Hesterberg R, Lorenz W, Schmidt U, Crombach M, Stahlknecht
CD (1985) Inhibition of human and canine diamine oxidase by drugs
used in an intensive care unit: relevance for clinical side effects? Agents
Actions 16: 91—94

Schoepp DD, Azzaro AJ (1981) Specificity of endogenous substrates for
types A and B monoamine oxidase in rat striatum. J Neurochem 36:
2025—2031

Schoepp DD, Azzaro AJ (1982) Role of type A and type B monoamine oxidase in the metabolism of released [^{3}H]dopamine from rat striatal slices.
Biochem Pharmacol 31: 2961—2968

Schoepp DD, Azzaro AJ (1983) Effects of intrastriatal kainic acid injection
on [^{3}H]dopamine metabolism in rat striatal slices: Evidence for post-
synaptic glial cell metabolism by both the type A and B forms of
monoamine oxidase. J Neurochem 40: 1340—1348

Seiler N (1981) Amide-bond-forming reactions of polyamines. In: Morris
DR, Marton LJ (eds) Polyamines in biology and medicine. Dekker, New
York, pp 127—149

Seiler N, Wiechmann M, Fischer HA, Werner G (1971) The incorporation of
putrescine carbon into gamma-aminobutyric acid in rat liver and brain
in vivo. Brain Res 28: 317—325

Shaff RE, Beaven MA (1976) Turnover and synthesis of diamine oxidase
(DAO) in rat tissues. Studies with heparin and cycloheximide. Biochem
Pharmacol 25: 1057—1062

Sourkes TL, Missala K (1981) Putrescine metabolism and the study of
diamine oxidase activity in vivo. Agents Actions 11: 20—27

Squires R (1972) Multiple forms of monoamine oxidase in intact mitochondria as characterized by selective inhibitors and thermal stability: a
comparison of eight mammalian species. In: Costa E, Sandler M (eds)
Monoamine oxidases—new vistas. Raven Press, New York (Adv biochem
psychopharmacol, vol 5, pp 355—370)

Squires RF, Lassen JB (1974) Inhibition of both A and B forms of MAO
required for production of characteristic behavioral syndrome in rats
after L-tryptophan loading. J Pharmacol 5: 96—CL 2

Stenström A, Arai Y, Oreland L (1985) Intra- and extraneuronal mono-
amineoxidase A and B activities after central axotomy (hemisection) on
rats. J Neural Transm 61: 105–113

Stjärne L (1975) Basic mechanisms and local feedback control of secretion of
noradrenergic and cholinergic neurotransmitters. In: Iversen LL, Iversen SD, Snyder SH (eds) Handbook of psychopharmacology, vol 6. Plenum Press, New York, pp 179–233

Strolin-Benedetti M, Boucher T, Carlsson A, Fowler CJ (1983) Intestinal
metabolism of tyramine by both forms of monoamine oxidase in the
rat. Biochem Pharmacol 32: 47–52

Student AK, Edwards DJ (1977) Subcellular localization of types A and B
monoamine oxidase in rat brain. Biochem Pharmacol 26: 2337–2342

Suzuki O, Hattori H, Masakazu O, Katsumata Y (1980) Characteristics of
monoamine oxidase in mitochondria isolated from chick brain, liver,
kidney and heart. Biochem Pharmacol 29: 603–607

Suzuki O, Katsumata Y, Masakazu O (1982) Substrate specificity of type A
and type B monoamine oxidase. In: Kamijo K, Usdin E, Nagatsu T (eds)
Monoamine oxidase: basic and clinical frontiers. Excerpta Medica,
Amsterdam, pp 74–86

Suzuki O, Matsumoto T (1985) Normetanephrine and metanephrine oxidized by both types of monoamine oxidase. Experientia 41: 634–636

Tipton KF, Youdim MB, Spires IP (1972) Beef adrenal medulla monoamine
oxidase. Biochem Pharmacol 21: 2197–2204

Tipton KF, Houslay MD, Garrett NJ (1973) Allotopic properties of human
brain monoamine oxidase. Nature 246: 213–214

Tipton KF, Fowler CJ, Houslay MD (1982) Specificities of the two forms of
monoamine oxidase. In: Kamijo K, Usdin E, Nagatsu T (eds) Monoamine oxidase: basic and clinical frontiers. Excerpta Medica, Amsterdam, pp 87–99

Urwyler S, Von Wartburg JP (1980) Studies on the subcellular localization of
monoamine oxidase types A and B and its importance for the deamination of dopamine in the rat brain. Biochem Pharmacol 29: 3067–3073

Van Der Krogt JA, Koot-Gronsveld E, Van Den Berg CJ (1983) Localization
of rat striatal monoamine oxidase activities towards dopamine, serotonin and kynuramine by gradient centrifugation and nigro-striatal
lesions. Life Sci 33: 615–623

Von Euler US (1972) Synthesis, uptake and storage of catecholamines in
adrenergic nerves. The effects of drugs. In: Eichler O, Fark A, Herken H,
Welch AD (eds) Handb exp pharmacol, vol 33. Springer, Berlin Heidel-
berg New York, pp 186–230

Waldmeier PC, Antonin KH, Feldtrauer JJ, Grunenwald C, Paul E, Lauber J,
Bieck P (1985) Urinary excretion of O-methylated catecholamines, tyramine, and phenylethylamine in human volunteers treated with tranylcypromine and CGP 11305 A. Eur J Clin Pharmacol 25: 361–368

Waldmeier PC, Delini-Stula A, Maître L (1976) Preferential deamination of
dopamine by an A type monoamine oxidase in rat brain. Naunyn-
Schmiedeberg's Arch Pharmacol 292: 9–14

White HL, Glassman AT (1977) Multiple binding sites of human brain and liver monoamine oxidase: substrate specificities, selective inhibitions, and attempts to separate enzyme forms. J Neurochem 29: 987—997

Wolf WA, Youdim MB, Kuhn DM (1985) Does brain 5-HIAA indicate serotonin release or monoamine oxidase activity? Eur J Pharmacol 109: 381—387

Yang HY, Neff NH (1974) The monoamine oxidases of brain: Selective inhibition with drugs and the consequences for the metabolism of the biogenic amines. J Pharmacol Exper Ther 189: 733—740

Youdim MB (1983) *In vivo,* noradrenaline is a substrate for rat brain monoamine oxidase A and B. Br J Pharmac 79: 477—480

Yu PH (1979) Effect of lipid depletion on type-A and type-B monoamine oxidase of rat heart and bovine liver mitochondria. In: Singer TP, Von Korff RW, Murphy DL (eds) Monoamine oxidase: Structure, function and altered functions. Academic Press, New York, pp 233—244

Zeller EA (1963) Diamine oxidases. In: Boyer PD, Lardy H, Myrbäck K (eds) The enzymes, vol 8. Academic Press, New York, pp 313—335

Zeller EA, Arora KL (1979) On the role of hydroxylic and N-methyl groups in the interaction of phenylethylamines with monoamine oxidase types A and B. In: Usdin E, Kopin FJ, Barchas JD (eds) Catecholamines: basic and clinical frontiers, vol 1. Pergamon, New York, pp 195—197

Author's address: Dr. P. C. Waldmeier, Research Department, Pharmaceuticals Division, Ciba-Geigy Ltd., CH-4002 Basle, Switzerland.

J Neural Transm (1987) [Suppl] 23: 73—89

Processing of MPTP by monoamine oxidases: implications for molecular toxicology

A. J. Trevor[1, 2], T. P. Singer[1, 3, 4],
R. R. Ramsay[4], and N. Castagnoli, Jr.[1, 2]

[1] Department of Pharmaceutical Chemistry,
[2] Department of Pharmacology,
[3] Department of Biochemistry and Biophysics, University of California,
San Francisco, and
[4] Molecular Biology Division, Veterans Administration Medical Center,
San Francisco, California, U.S.A.

Summary

MPTP (1-methyl-4-phenyl-1, 2, 3, 6-tetrahydropyridine), a selective nigrostriatal neurotoxin, is bioactivated by MAO-B (and less effectively by MAO-A) to 2, 3-MPDP$^+$ and this intermediate undergoes further oxidation to MPP$^+$, partly through the activity of MAO forms. MPTP and its two primary metabolites are competitive inhibitors of both A and B forms of MAO. MPTP and 2, 3-MPDP$^+$ are also mechanism-based inactivators of both forms of the enzyme. A catalytic mechanism, involving the formation of radical intermediates, is proposed for the MAO-mediated oxidation of MPTP. Post-oxidation biochemical sequelae, possibly involved in the expression of neurotoxicity, include the active accumulation of MPP$^+$ via dopamine reuptake systems, the energy-driven uptake of MPP$^+$ by mitochondria and the inhibition of NADH dehydrogenase by pyridine derivatives. A scheme linking these events as steps in the molecular mechanism of action of MPTP is proposed and discussed in terms of the selective toxicity of the neurotoxin towards nigrostriatal cells.

Introduction

MPTP (1-methyl-4-phenyl-1, 2, 3, 6-tetrahydropyridine), the causative agent of parkinsonian symptoms in drug abusers who have

self-administered certain samples of an illicit narcotic-analgesic
(Langston *et al.,* 1983), elicits similar symptoms in experimental
animals and causes selective degeneration of dopamine-containing
nigrostriatal cells (Burns *et al.,* 1983). Considerations of the molecular
structure of MPTP give no immediate indications of its potential
chemical reactivity, nor why its cytotoxic actions should be selective.
However, based on our previous studies on the oxidative metabolism
of psychoactive tertiary amines it appeared possible that MPTP
might be bioactivated *in vivo* to unleash its potential toxicity. Verifi-
cation of this possibility came from observations that MPTP was oxidized by rat brain mitochondrial fractions to 2, 3-MPDP$^+$ and MPP$^+$
(Fig. 1), its 2-electron and 4-electron oxidation products, respectively
(Chiba *et al.,* 1984), the latter being the metabolite isolated from the
substantia nigra of MPTP-treated primates (Markey *et al.,* 1983).
Further, pargyline and deprenil blocked MPTP oxidation by brain
mitochondria (Chiba *et al.,* 1984) and the prior administration of
such MAO inhibitors prevented the development of MPTP neurotoxicity in animals (Heikkila *et al.,* 1984; Langston *et al.,* 1984). Direct
proof of the involvement of MAO in the metabolic activation of
MPTP was obtained from experiments with purified forms of the
enzyme, resulting in kinetic characterization of MPTP and its
metabolites as substrates, competitive inhibitors and mechanism-
based inactivators of MAO-A and B (Salach *et al.,* 1984; Singer *et al.,*
1985; Singer *et al.,* 1986).

These studies are of considerable interest from the standpoint of
the biochemistry and pharmacology of MAO, but they do not provide a direct answer to the question of why the toxic effects of MPTP
are so highly selective for the dopaminergic neurons in the substantia
nigra. While MPP$^+$ accumulates in such cells, MPTP is oxidized
mainly by MAO-B which is localized in glia and serotonin-con-
taining neurons (Westlund *et al.,* 1985) and these cells are unaffected
by the neurotoxin. Explanations for this apparent anomaly have

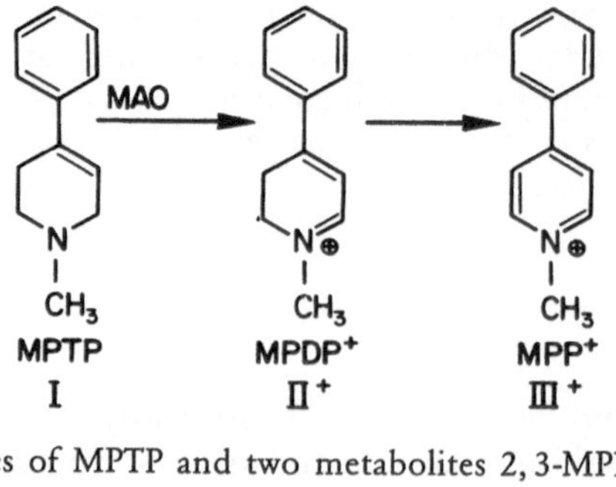

Fig. 1. Structures of MPTP and two metabolites 2, 3-MPDP$^+$ and MPP$^+$

been partly afforded by the demonstration that MPP$^+$ is an effective substrate for synaptosomal dopamine uptake systems (Javitch et al., 1985; Chiba et al., 1985). In terms of the possible mechanisms involved in the expression of the neurotoxicity of MPTP, biochemical studies of its oxidation products have shown that MPP$^+$ interferes with mitochondrial respiration (Nicklas et al., 1985) via inhibition of NADH dehydrogenase (Ramsay et al., 1986 a) and that it is accumulated in mitochondria by a novel energy-dependent uptake mechanism (Ramsay et al., 1986 b).

In this paper the MAO-mediated oxidation of MPTP and mechanism-based inactivation of the enzyme forms are described and a possible catalytic mechanism for the oxidation reactions is proposed. The biochemical properties of the oxidation product MPP$^+$ are reviewed in terms of a sequence of critical events that may lead to cytotoxicity. The apparent selective toxicity of MPTP for dopamine-containing nigrostriatal cells is discussed.

MAO and the metabolic activation of MPTP

Highly purified preparation of MAO-B from beef liver and MAO-A from human placenta have been used to determine the oxidation rates of MPTP, in comparison with preferred substrates for the two enzyme forms (Salach et al., 1984; Singer et al., 1985). From true initial rate determinations the turnover number of the B enzyme with MPTP was found to be approximately 200, compared with 530 for benzylamine, the fastest known substrate for this enzyme form and the K_m values were similar. With MAO-A, turnover numbers for MPTP and kynuramine were calculated to be 14 and 120, respectively. The rapid oxidation of MPTP, a tertiary amine, by MAO-B and even the substantial rate by the A form of the enzyme were unexpected and prompted our suggestion that a reexamination of the substrate specificity of both forms of MAO is needed, particularly with regard to tertiary amine xenobiotics. The 2-electron oxidation product, 2, 3-MPDP$^+$, was also oxidized by MAO with a turnover number of 6 with both forms of the enzyme, forming MPP$^+$. Although 2, 3-MPDP$^+$ may undergo disproportionation to form MPTP and MPP$^+$ (Peterson et al., 1985) and may also be oxidized to MPP$^+$ by endogenous compounds including neuromelanin (Wu et al., 1986), MAO forms may also play a significant role in the intracellular generation of MPP$^+$. No evidence was found that MPP$^+$ undergoes oxidation through the mediation of MAO and it is clearly not a substrate for either form of the enzyme.

MPTP, as well as its 2-electron and 4-electron oxidation products, are effective competitive inhibitors of both types of MAO (Singer *et al.*, 1985), with the A form being more sensitive. The K_i value of 3 μM for MPP$^+$ and 6 μM for 2, 3-MPDP$^+$ in the case of MAO-A, suggest that intracellular accumulation of these metabolites of MPTP would be likely to exert inhibitory effects on this form of the enzyme *in vivo*.

The incubation of MAO-A or MAO-B with either MPTP or 2, 3-MPDP$^+$ (but not MPP$^+$) led to a time-dependent inactivation of enzyme activity (Fig. 2). These effects followed pseudo first order kinetics, were protected against by co-incubation with normal substrates and were only partially reversible by prolonged dialysis, indicative of a mechanism-based or "suicide" inactivation of MAO (Singer *et al.*, 1986). Incubation of ^{3}H-MPTP with MAO-B resulted in a time-dependent association of radiolabel with the protein, reaching a maximum of 5 mol of radioactive product bound/mol of enzyme. This ratio may reflect the high partition coefficient associated with

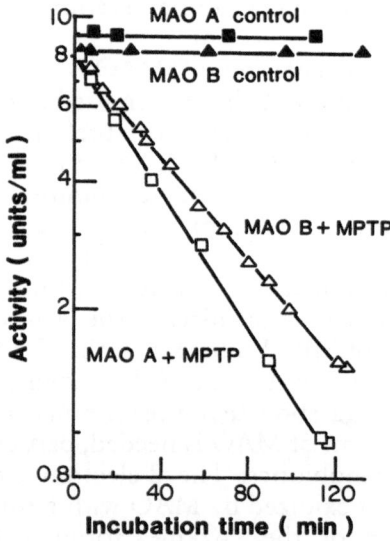

Fig. 2. Inactivation of MAO forms by MPTP. Purified MAO-A from human placental mitochondria was incubated at 30 °C with 5 mM MPTP in a standard assay mixture ($\square$); control samples were similarly incubated without MPTP ($\blacksquare$). Samples of 10 μl were periodically removed and added to 0.99 ml of assay mixture containing 1 mM kynuramine for strict initial-rate spectrophotometric measurement of enzyme activity. Parallel experiments were conducted with purified MAO-B from bovine liver mitochondria in the absence ($\blacktriangle$), or presence ($\triangle$), of 2 mM MPTP and enzyme activity assayed periodically using 3.3 mM benzylamine

the inactivation process, as well as the possible reaction of electrophilic intermediates generated at the catalytic site with more distant nucleophilic functionalities of the enzyme. The molecular identity of the reactive intermediate(s) and its binding site(s) remain to be determined. Covalently-bound flavin groups do not appear to be involved in the inactivation process, since the flavin peptides recovered from proteolysis of the inactivated enzyme display spectral properties typical of cysteinylflavin peptides, whereas adducts formed by covalent interactions at N-5 or C-4 a of the flavin do not. It is possible that reactive species generated from the oxidation of MPTP (or 2, 3-MPDP$^+$) undergo nucleophilic attack by the -SH group of the active site of MAO, in an analogous manner to adduct formation suggested for mechanism-based inactivation of the enzyme by phenylcyclopropylamine (Silverman and Yamasaki, 1984). Such adducts dissociate slowly on prolonged dialysis and in this regard it is of interest that the activity of MPTP-inactivated MAO could be partially restored by such treatment.

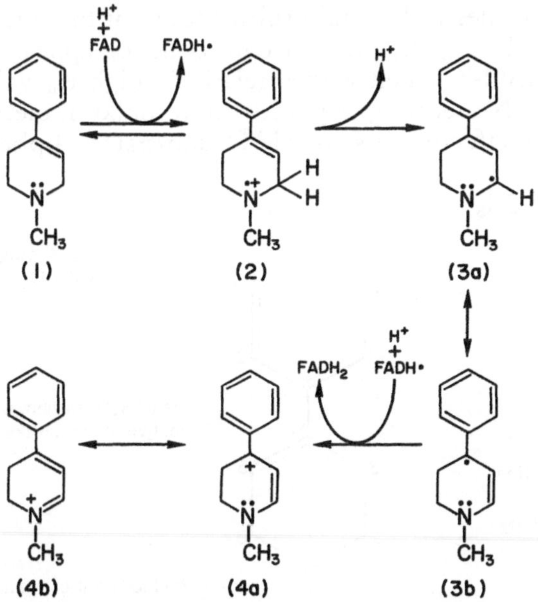

Fig. 3. Possible catalytic mechanism for the oxidation of MPTP by MAO-B. MPTP *(1)* generates the aminium radical cation *(2)* on transfer of one electron to FAD. This is followed by abstraction of a proton resulting in formation of the resonance-stabilized species *(3 a, b)*, which upon transfer of a second electron to FADH$^+$ would yield 2, 3-MPDP$^+$ *(4 b)*

The catalytic mechanism of MAO is not fully elucidated. Recent studies with "suicide" substrates including benzylcyclopropylamines (Silverman and Yamasaki, 1984) and 1-phenylcyclobutylamine (Silverman and Zieske, 1986) have led to the suggestion of a mechanism involving an initial 1-electron transfer from the substrate to FAD, followed by loss of a proton and a second 1-electron transfer to the flavin radical intermediate. When applied to MPTP (Fig. 3) loss of one electron to FAD would generate the aminium radical cation *2*. Abstraction of a proton would result in formation of the resonance stabilized radical species *3 a* (— — —) *3 b,* which upon loss of a second electron would yield MPDP$^+$ directly.

This proposed catalytic mechanism for the oxidation of MPTP by MAO remains quite speculative. It is clear that the molecule is an exceptionally good substrate for MAO, especially the B form of the enzyme. The structural requirements that determine substrate specificity for the MAO forms in the tetrahydropyridine series of molecules appear to be quite stringent (Heikkila *et al.*, 1985; Brossi *et al.*, 1986). Many congeners of MPTP are oxidized by the enzyme forms at lower rates, or are inactive as substrates (Fig. 4). These include molecules with certain substitutions at the prime 4 position of the phenyl ring; the prime 4 chloro analog appears to be a substrate for MAO-B, albeit with a relatively high K_m value. Replacement of the N-methyl group with more bulky moieties results in decreases in effectiveness as MAO substrates; 4-phenyl-1, 2, 3, 6-

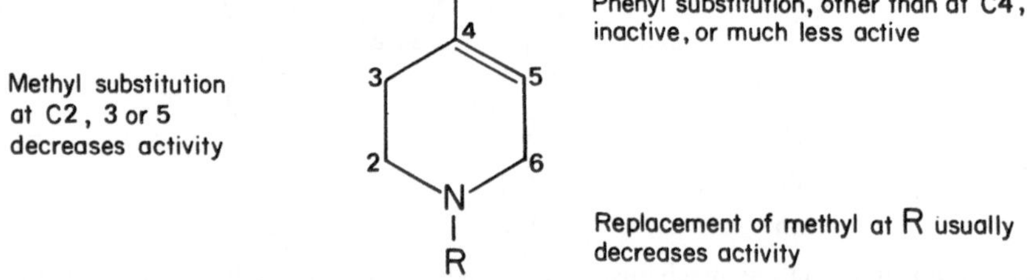

MPTP : R = CH$_3$

Substitution at 4' (e.g. Cl, OH) usually decreases activity

Phenyl substitution, other than at C4, inactive, or much less active

Methyl substitution at C2, 3 or 5 decreases activity

Replacement of methyl at R usually decreases activity

Fig. 4. Substituted phenyl-tetrahydropyridines as substrates for MAO-B. Most substituted phenyl-tetrahydropyridines that have been assessed as substrates for MAO-B appear to be less effective than MPTP, or totally inactive. 2'-methyl-MPTP is a notable exception and is an effective neurotoxin

tetrahydropyridine (the N-demethylated metabolite of MPTP) does appear to be a substrate for MAO-B, but with a higher K_m value than MPTP. Methyl substitutions at various positions (C 2, 3 and 5) in the tetrahydropyridine ring appear to decrease substrate effectiveness based on studies with purified MAO-B. Very few structural analogs of MPTP have been reported to be neurotoxins. One notable exception is 2'-methyl-MPTP, which in rodent species is more neurotoxic than MPTP (Youngster et al., 1986). In rat brain mitochondrial preparations 2'-methyl-MPTP is oxidized more rapidly than MPTP, suggesting that the compound is also an effective substrate for MAO. However, prior administration of MAO-B inhibitors do not effectively block the development of the neurotoxic effects of 2'-methyl-MPTP in vivo. Proof of the effectiveness of the compound to act as a substrate for MAO, awaits the investigation of its oxidation with purified forms of the enzyme.

MPP$^+$ and mitochondrial functions

Although is has not been ruled out that 2, 3-MPDP$^+$, or a labile reactive oxidation product formed from this molecule, is involved in the initiation of the neurotoxic actions of MPTP, most attention in terms of the attempted elucidation of the mechanisms underlying toxicity, has focussed on the 4-electron oxidation product, MPP$^+$. Reasons for this include its active accumulation in nigrostriatal cells, the protection of dopamine-depleting effects of MPTP by dopamine reuptake blockers and the direct demonstration of its cytotoxic actions in isolated cell systems (DiMonte et al., 1986). Based on inhibition of the oxidation of NAD$^+$-linked substrates in rat liver and brain mitochondrial preparations in State 3 by moderate concentrations (0.5 mM) of MPP$^+$, it was proposed that the compound blocks mitochondrial NADH oxidation resulting in ATP depletion and cell death (Nicklas et al., 1985). Since charged molecules are generally thought not to cross the inner membrane of mitochondria freely, these reported effects of MPP$^+$ on mitochondrial respiration were unexpected. However, in systematic studies of the effects of MPP$^+$ (0.5 mM) on mitochondrial function (Ramsay et al., 1986 a) inhibition of the oxidation of NAD$^+$-linked substrates was confirmed using intact liver mitochondria (Fig. 5) while succinate oxidation was unaffected. Collapse of the electrical gradient upon addition of uncouplers resulted in a partial reversal of the inhibitory action of MPP$^+$, suggesting the possibility of its concentration in mitochondria by an energy-dependent mechanism. After brief sonication of mito-

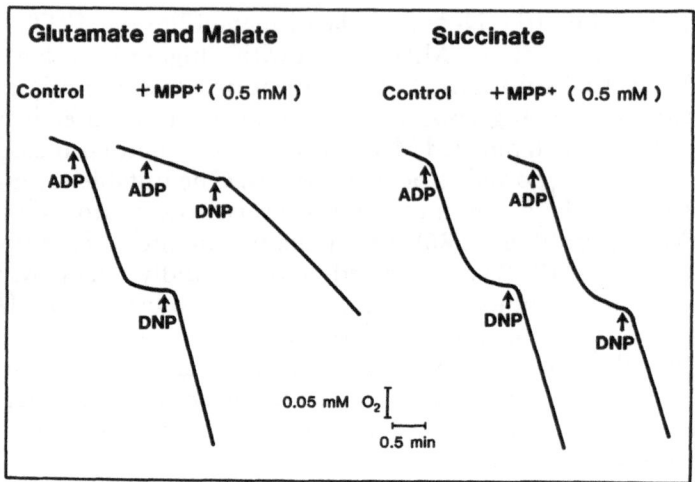

Fig. 5. Effects of MPP⁺ on respiration of rat liver mitochondria. Respiration of intact mitochondria from rat liver (1.9 mg/ml) at 25 °C was measured polarographically following 5 min preincubation in the presence of substrates with or without 0.5 mM MPP⁺. This concentration of MPP⁺ inhibited glutamate (or malate) oxidation, and this effect was partly reversed by addition of 2, 4-DNP (39 μM). In contrast, succinate oxidation in State 3 or after uncoupling was unaffected by MPP⁺

chondria to invert the inner membranes 0.5 mM MPP⁺ no longer inhibited NADH oxidation. A similar loss of sensitivity to inhibition was observed in mitochondrial ETP preparation (respiratory chain minus the coupling factors) and Complex I, its initial segment. However, such preparations, used customarily to study inhibitors of NADH oxidation, could be inhibited by high concentrations of MPP⁺ (Fig. 6). These data compare the inhibition of NADH oxidation of ETP preparations from beef heart mitochondria by several substituted pyridines (Ramsay *et al.*, 1986 b) and reveal that MPP⁺ is considerably less effective than 4-phenylpyridine and MPTP itself. As seen in Table 1 these inhibitory effects were freely reversible on dilution and appear to require the continued presence of the compounds to exert their actions on NADH-linked oxidation.

Since steady-state concentrations of MPP⁺ in nigrostriatal cells of MPTP-treated animals are lower by several orders of magnitude than those used in our *in-vitro* experiments, the possible relevance of the observed inhibitory effects of MPP⁺ on mitochondrial NADH oxidation could be questioned. However, we have recently provided evidence for an energy-dependent uptake system for MPP⁺ in mitochondria, which rapidly concentrates the compound at micromolar external concentrations (Ramsay *et al.*, 1986 c). This novel mito-

Table 1. Reversibility of inhibition of NADH-linked oxidation
by substituted pyridines

Inhibition	Initial concentration (mM)	Inhibition (%)	Concentration after dilution (mM)	Inhibition (%)
MPTP	7.0	96	0.55	0
MPP+	20	93	0.50	6
4-phenylpyridine	1.1	93	0.027	0
N-methylphenylpiperidone	3	75	0.075	0

Inner mitochondrial membrane (ETP) preparations were incubated in buffered medium (pH 7.6) for 5 min at 30 °C at the initial concentrations of inhibitors indicated and were then assayed spectrophotometrically for initial rates of NADH oxidation with and without a 40-fold dilution.

chondrial uptake system, for which there is no known endogenous substrate, requires the presence of oxidizable substrates, is temperature-dependent and is inhibited by uncouplers. The driving force for the accumulation of MPP+ is the membrane electrical potential (Fig. 7). The rate of MPP+ uptake with increasing K^+ concentration in the presence of valinomycin parallels the estimated change in membrane potential. Such conditions have little or no effect on the proton gradient. Of particular interest is the fact that the uptake system

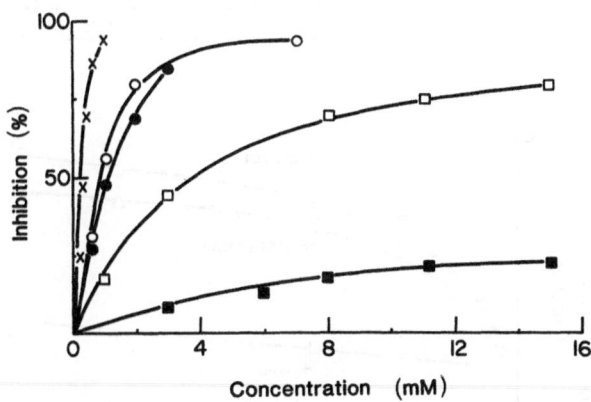

Fig. 6. Effects of substituted pyridines on NADH oxidation of inner membrane preparations from mitochondria (ETP). ETP preparations from beef heart mitochondria were incubated 5 min at 25 °C in phosphate buffer (pH 7.6) with the indicated concentrations of substituted pyridines. Samples were taken for the spectrophotometric determination of initial rates of oxidation of NADH. Symbols represent: X——X 4-phenylpyridine; O——O MPTP; ●——● N-methyl-4-phenylpyridinone; ■——■ N-methylpyridine, □——□ MPP+

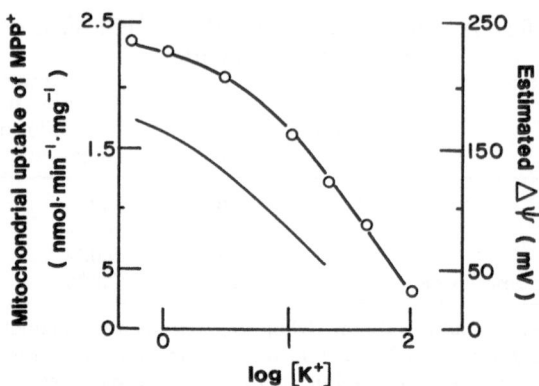

Fig. 7. Dependence on membrane potential of ^{3}H-MPP$^+$ uptake by mitochondria. The uptake of ^{3}H-MPP$^+$ by intact mitochondria from rat liver was measured for 1 min during incubation at 25 °C in buffered media (pH 7.4) containing pyruvate, malate and valinomycin (0.8 μg/ml) at various K$^+$ concentrations (O). The solid line shows the calculated change in membrane potential with increasing K$^+$ based on published data

is present in isolated brain mitochondria (Ramsay *et al.*, 1986 d). The uptake of ^{3}H-MPP$^+$ by mitochondria isolated from rat mid-brain regions is shown in Fig. 8. MPP$^+$ mitochondrial uptake in such preparations can be distinguished from its binding to synaptosomal components by the minimal effects of dopamine, or mazindol, on its accumulation. The energy-dependent uptake and accumulation of MPP$^+$ occurs against its concentration gradient. This is shown by the

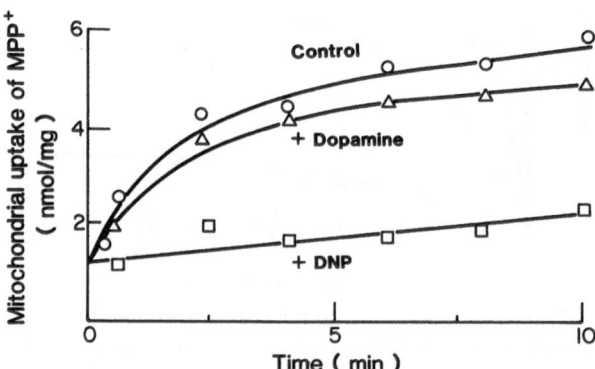

Fig. 8. Uptake of ^{3}H-MPP$^+$ by intact mitochondria from rat midbrain was determined during incubation in buffered media (pH 7.4) containing glutamate and malate (O). The uncoupling agent 2, 4-DNP ($\square$) at 6×10^{-6} M inhibited ^{3}H-MPP$^+$ uptake in brain mitochondria, but 10^{-5} M dopamine ($\triangle$) had marginal effects

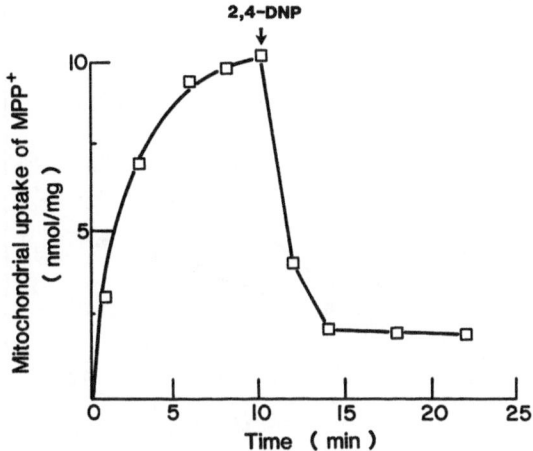

Fig. 9. Efflux of ^{3}H-MPP$^+$ from mitochondria. Uptake of ^{3}H-MPP$^+$ was measured in intact mitochondria from rat liver during incubation in buffered medium (pH 7.4) with malate. At 10 min, 2, 4-DNP (60 μM) was added. Note the rapid efflux of radiolabelled MPP$^+$ on addition of the uncoupling agent

data in Fig. 9 where addition of an uncoupler in the course of uptake caused immediate efflux of the accumulated MPP$^+$.

These experiments reconcile the observations that in intact mitochondria the inhibitory effects of MPP$^+$ on NAD$^+$-linked oxidation occurs at lower concentrations than those required to block NADH oxidation in mitochondrial membrane preparations such as ETP or Complex I. Their possible relevance to the neurotoxic actions of MPP$^+$ will be discussed below.

Experiments designed to reveal the precise site of action of MPP$^+$ on mitochondrial NADH oxidation have excluded the point of entry of pyridine-nucleotide generating reducing equivalents into the electron-transport chain. MPP$^+$ exerts no inhibitory effects on NADH-ferricyanide activity in ETP preparations (Fig. 10). Ferricyanide accepts electrons from the low potential Fe-S cluster 1 of NADH dehydrogenase and inhibitory actions at this site are readily detectable under normal circumstances. Neither have low-temperature EPR studies of the possible effects of MPP$^+$ on NADH dehydrogenase under conditions of reductive titration revealed spectral evidence for inhibitory effects on any of the five sequential Fe-S clusters (Singer *et al.*, 1986 b). Since succinate oxidation is unaffected by MPP$^+$, the most plausible site of its inhibitory effects on mitochondrial function appears to be between the high potential Fe-S cluster of NADH dehydrogenase and coenzyme Q (Fig. 11).

A. J. Trevor *et al.*

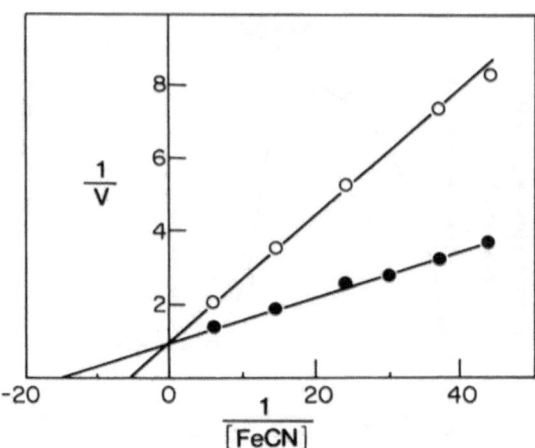

Fig. 10. Effects of substituted pyridines on NADH-ferricyanide activity of ETP preparations. Beef heart ETP preparations were incubated for 5 min at 30°C in triethanolamine buffer (pH 7.8) with ferricyanide and absorbance changes measured at 420 nm after addition of NADH. Data are shown as double reciprocal plots of initial rate of electron transfer versus ferricyanide concentration (reciprocal $\times 10^{-2}$ M). Symbols: O——O; control; ●——● 5 mM MPP$^+$

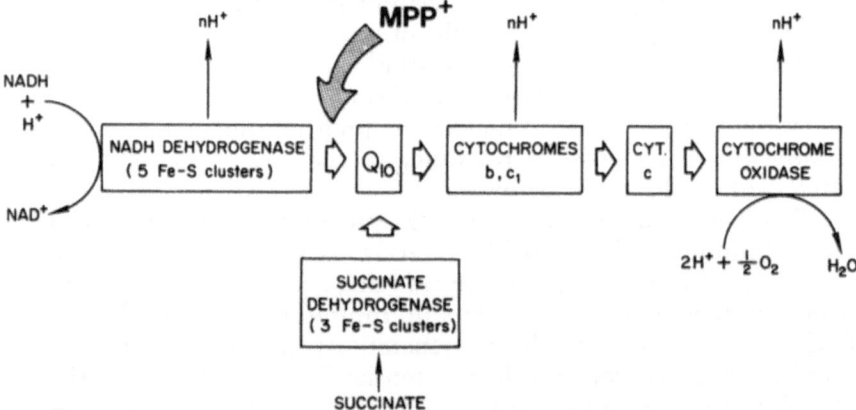

Fig. 11. Putative site of action of MPP$^+$ to inhibit mitochondrial respiration. The energy-driven accumulation of MPP$^+$ by mitochondria results in its concentration at the inner mitochondrial membrane. While the oxidation of NADH-linked substrates is inhibited by MPP$^+$, oxidation of succinate is not affected. MPP$^+$ does not inhibit electron transfer between the low potential Fe-S cluster of NADH dehydrogenase and ferricyanide and no spectral evidence has been obtained for effects on higher potential Fe-S clusters of the enzyme. Based on these observations MPP$^+$ appears to inhibit electron transfer between the high potential Fe-S cluster of NADH dehydrogenase and coenzyme Q

From these experiments it is possible to describe a plausible sequence of biochemical events which may explain nigrostriatal cytotoxicity of MPTP, with consequent depletion of dopamine and expression of parkinsonian symptoms. MPTP, by virtue of its high lipophilicity, has ready access to the CNS from the vascular system. In the brain the protoxin undergoes bioactivation mediated by MAO-B, presumably in cells other than the nigrostriatal dopaminergic neurons. By mechanisms yet to be explained, the potentially reactive oxidation products $2,3$-MPDP$^+$ and MPP$^+$ gain access to the extracellular space. The pyridinium oxidation product MPP$^+$ is a good substrate for dopamine uptake systems, through the mediation of which it is accumulated in nigrostriatal cells. Such selective neuronal uptake may be a critical biochemical event in that it could provide an intracellular concentration of MPP$^+$ favorable to the energy-dependent mitochondrial uptake system. Concentrations of 10 to 20 mM MPP$^+$ which are inhibitory to NADH oxidation can be reached in mitochondria during incubation in media containing 30–50 μM concentrations of MPP$^+$ and these are in the range found in brain regions containing nigrostriatal cells following MPTP treatment of rodents *in vivo*. The inhibition of mitochondrial NADH dehydrogenase resulting from MPP$^+$ uptake could lead to a cessation of oxidative phosphorylation and subsequent cytotoxic effects since central nervous system neurons are highly dependent on aerobic ATP synthesis for the maintenance of their cellular integrity. Since the inhibitory actions of MPP$^+$ are reversible, blockade of NADH oxidation would necessitate the continuous presence of the molecule at its site of action on the inner mitochondrial membrane. In this regard MPP$^+$ accumulates selectively in specific brain regions that include those containing nigrostriatal neurons, following administration of MPTP.

Selectivity of the cytotoxic effects of MPTP

The scheme of biochemical events proposed above leading to the expression of the cytotoxic effects of MPTP has a certain plausibility, but certain difficulties arise when one attempts a completely satisfactory explanation for the selectivity of the neurotoxin towards nigrostriatal cells. Questions that need resolution include:

1. Why are the neurons of other CNS dopaminergic pathways less sensitive to the neurotoxic effects of MPTP? Does MPP$^+$ have more limited access to such cells? Are such cells "protected" by endogenous mechanisms, or by adjacent cell types, not available to

nigrostriatal neurons. Is it possible that the nigrostriatal cells have characteristic biochemical properties that facilitate expression of the neurotoxicity of MPTP?

2. Why are MAO-B containing neurons and glia apparently resistant to the actions of MPTP when their mitochondrial enzymes presumably generate potentially toxic oxidation products? Do these cells have mechanisms for the active extrusion of 2, 3-MPDP$^+$ or MPP$^+$? Is mechanism-based inactivation of MAO in such cells self-limiting in the bioactivation of MPTP?

3. How do we explain the apparent resistance of the cells of the many peripheral tissues in the body that contain MAO, often in high activity, and which have the potential to generate oxidation products of MPTP via such enzymes? In the case of isolated hepatocytes, MPTP and its oxidation products, MPDP$^+$ and MPP$^+$ are all cytotoxic (DiMonte *et al.*, 1986). Do alternative pathways of oxidative metabolism of MPTP, such as those mediated by cytochrome P-450 and flavin monooxygenase (Weissman *et al.*, 1985), leading to the formation of inactive metabolites, play a protective role? Is the cell-regenerative capability of extracerebral tissues an adequate explanation for the limited evidence for peripheral cytotoxic effects of MPTP *in vivo*?

Clearly, a number of difficulties have yet to be overcome in the development of a rational explanation for the selective neurotoxicity of MPTP in term of the precise molecular events involved. The first critical event, involving a requirement for MAO-mediated bioactivation of MPTP, appears to be generally accepted. It has now become more widely recognized that brain MAO forms are not only involved in the pathways of oxidative metabolism of biogenic amine neurotransmitters but may also play a role in the metabolism of xenobiotics. Irrespective of the possible relationship between MPTP-induced nigrostriatal toxicity and idiopathic Parkinson's Disease, research interest in the neurotoxin has resulted in important advances in our knowledge of neurobiology and biochemistry. Not least, it has clearly demonstrated the exquisite selectivity with which an apparently innocuous molecule can be converted into a devastating neurotoxin.

Acknowledgements

This work was supported in part by NIDA research grant DA 03405, NINCD research grant NS 23066, NIH Program Project HL 16251, NSF Grant DM 846967 and the Veterans Administration.

References

Brossi A, Gessner WP, Fritz RR, Bembenek ME, Abell CW (1986) Interaction of monoamine oxidase B with analogues of 1-methyl-4-phenyl-1, 2, 3, 6-tetrahydropyridine derived from prodine-type analgesics. J Med Chem 29: 445–448

Burns RS, Chiueh CC, Markey SP, Ebert MG, Jacobowitz DM, Kopin IJ (1983) A primate model of parkinsonism: Selective destruction of dopaminergic neurons in the pars compacta of the substantia nigra by N-methyl-4-phenyl-1, 2, 3, 6-tetrahydropyridine. Proc Natl Acad Sci USA 80: 4546–4550

Chiba K, Trevor A, Castagnoli N (1984) Metabolism of the neurotoxic tertiary amine, MPTP, by brain monoamine oxidase. Biochem Biophys Res Comm 120: 574–578

Chiba K, Trevor AJ, Castagnoli N (1985) Active uptake of MPP$^+$, a metabolite of MPTP, by brain synaptosomes. Biochem Biophys Res Commun 128: 1228–1232

DiMonte D, Jewell SA, Ekstrom G, Sandy MS, Smith MT (1986) 1-methyl-4-phenyl-1, 2, 3, 6-tetrahydropyridine (MPTP) and 1-methyl-4-phenylpyridine (MPP$^+$) cause rapid ATP depletion in isolated hepatocytes. Biochem Biophys Res Commun 137: 310–319

Heikkila RE, Manzino L, Cabbat FS, Duvoisin RC (1984) Protection against dopaminergic neurotoxicity of 1-methyl-4-phenyl-1, 2, 3, 6-tetrahydropyridine by monoamine oxidase inhibitors. Nature 311: 467–469

Heikkila RE, Manzino L, Cabbat FS, Duvoisin RC (1985) Studies on the oxidation of the dopaminergic neurotoxin 1-methyl-4-phenyl-1, 2, 3, 6-tetrahydropyridine by monoamine oxidase B. J Neurochem 45: 1049–1054

Javitch JA, D'Amato RJ, Strittmatter SM, Snyder SH (1985) Parkinsonism-inducing neurotoxin N-methyl-4-phenyl-1, 2, 3, 6-tetrahydropyridine: Uptake of the metabolite N-methyl-4-phenylpyridine by dopamine neurons explains selective toxicity. Proc Natl Acad Sci USA 82: 2173–2177

Langston JW, Ballard P, Tetrud JW, Irwin I (1983) Chronic Parkinsonism in humans due to a product of meperidine-analog synthesis. Science 219: 979–980

Langston JW, Irwin I, Langston EB, Forno LS (1984) Pargyline prevents MPTP-induced parkinsonism in primates. Science 225: 1480–1482

Markey SP, Johannessen JN, Chiueh CC, Burns RS, Herkenham MA (1984) Intraneural generation of a pyridinium metabolite may cause drug-induced parkinsonism. Nature 311: 464–467

Nicklas WJ, Vyas I, Heikkila RE (1985) Inhibition of NADH-linked oxidation in brain mitochondria by 1-methyl-4-phenyl-pyridine, a metabolite of the neurotoxin, 1-methyl-4-phenyl-1, 2, 5, 6-tetrahydropyridine. Life Sci 36: 2503–2508

Peterson LA, Caldera PS, Trevor A, Chiba K, Castagnoli N (1985) Studies on the 1-methyl-4-phenyl-2, 3-dihydropyridinium species, 2, 3-MPDP⁺, the monoamine oxidase catalyzed oxidation product of the nigrostriatal toxin 1-methyl-4-phenyl-1, 2, 3, 5-tetrahydropyridine (MPTP). J Med Chem 28: 1432–1436

Ramsay RR, Salach JI, Singer TP (1986 a) Uptake of the neurotoxin 1-methyl-4-phenylpyridine (MPP⁺) by mitochondria and its relation ot the inhibition of the mitochondrial oxidation of NAD⁺-linked substrates by MPP⁺. Biochem Biophys Res Commun 134: 743–748

Ramsay RR, Salach JI, Dadgar J, Singer TP (1986 b) Inhibition of mitochondrial NADH dehydrogenase by pyridine derivatives and its possible relation to experimental and idiopathic Parkinsonism. Biochem Biophys Res Commun 135: 269–275

Ramsay RR, Singer TP (1986 c) Energy-dependent uptake of N-methyl-4-phenylpyridinium, the neurotoxic metabolite of 1-methyl-4-phenyl-1, 2, 3, 6-tetrahydropyridine, by mitochondria. J Biol Chem 261: 7585–7587

Ramsay RR, Dadgar J, Trevor A, Singer TP (1986) Energy-driven uptake N-methyl-4-phenylpyridine by brain mitochondria mediates the neurotoxicity of MPTP. Life Sci 39: 581–588

Salach JI, Singer TP, Castagnoli N, Trevor AJ (1984) Oxidation of the neurotoxic amine 1-methyl-4-phenyl-1, 2, 3, 6-tetrahydropyridine (MPTP) by monoamine oxidases A and B and suicide inactivation of the enzymes by MPTP. Biochem Biophys Res Commun 125: 831–835

Silverman RB, Yamasaki RB (1984) Mechanism-based inactivation of mitochondrial monoamine oxidase by N-(1-methylcyclopropyl)benzylamine. Biochemistry 23: 1322–1332

Silverman RB, Zieske PA (1986) 1-phenylcyclobutylamine the first in a new class of monoamine oxidase inactivators. Further evidence for a radical intermediate. Biochemistry 25: 341–346

Singer TP, Salach JI, Crabtree D (1985) Reversible inhibition and mechanism-based irreversible inactivation of monoamine oxidases by 1-methyl-4-phenyl-1, 2, 3, 6-tetrahydropyridine (MPTP). Biochem Biophys Res Commun 127: 707–712

Singer TP, Salach JI, Castagnoli N, Trevor AJ (1986) Interactions of the neurotoxic amine 1-methyl-4-phenyl-1, 2, 3, 6-tetrahydropyridine with monoamine oxidases. Biochem J 235: 785–789

Weissman J, Trevor A, Chiba K, Peterson L, Caldera P, Castagnoli N (1985) Metabolism of the nigrostriatal toxin 1-methyl-4-phenyl-1, 2, 3, 6-tetrahydropyridine by liver homogenate fractions. J Med Chem 28: 997–1001

Westlund KN, Denney RM, Kochenperger LM, Rose RM, Abell CW (1985) District monoamine oxidase A and B populations in primate brain. Science 230: 181–183

Wu EY, Chiba K, Trevor AJ, Castagnoli N (1986) Interactions of the 1-methyl-4-phenyl-2, 3-dihydropyridinium species with synthetic dopamine-melanin. Life Sci 39: 1695–1700

Youngster SK, Duvoisin RC, Hess A, Sonsalla PK, Kindt MV, Heikkila
RE (1986) 1-methyl-4-(2'-methyl-phenyl)-1, 2, 3, 6-tetrahydropyridine
(2'-C 4₃-MPTP) is a more potent dopaminergic neurotoxin than MPTP
in mice. Eur J Pharmacol 122: 283–287

Authors' address: Dr. A. J. Trevor, Professor of Pharmacology and Pharma-
ceutical Chemistry, Department of Pharmacology, School of Medicine,
Medical Science Building, S 1210, University of California, 3rd and
Parnassus Streets, San Francisco, CA 94143, U.S.A.

J Neural Transm (1987) [Suppl] 23: 91–101

The functional coupling of neuronal and extraneuronal transport with intracellular monoamine oxidase*

U. Trendelenburg, L. Cassis, M. Grohmann, and A. Langeloh

Pharmakologisches Institut, Universität Würzburg, Würzburg, Federal Republic of Germany

Summary

"Metabolizing systems" are responsible for the quick inactivation of noradrenaline released from adrenergic nerve endings: a transport mechanism (uptake$_1$ or uptake$_2$) is arranged in series with the intracellular enzyme (monoamine oxidase, MAO; catechol-O-methyltransferase, COMT). In the perfused rat heart, k_{enzyme}-values were determined, *i.e.*, those rate constants which characterize the unsaturated intracellular enzymes. In the extraneuronal metabolizing system $k_{comt} > k_{mao}$ for noradrenaline and adrenaline, while rather similar rate constants were obtained for dopamine. However, for the neuronal deaminating system, k_{mao} is considerably higher than k_{mao} for the extraneuronal system. Second, in the rat vas deferens it is demonstrated that inhibition of neuronal MAO leads to very pronounced rises of the axoplasmic noradrenaline concentration—and this is again a reflection of the high activity of neuronal MAO. In a third series of experiments (with the rat vas deferens), the evidence indicates that the neuronal inward transport of substrates of MAO fails to saturate the enzyme. This is the functional consequence of the high activity of neuronal MAO. It is concluded that a) neuronal MAO activity is very high, and—as a consequence—b) axoplasmic noradrenaline levels are very low.

* The experiments reported here were supported by the Deutsche Forschungsgemeinschaft (SFB 176). — L. C. and A. L. were recipients of Humboldt-Foundation scholarships.

Metabolizing systems

Fig. 1 is a schematic representation of those "metabolizing systems" that ensure the quick inactivation of noradrenaline after its exocytotic release from adrenergic varicosities: as the transmitter is highly polar and nearly unable to diffuse through cell membranes, and as the metabolizing enzymes (monoamine oxidase, MAO, and catechol-O-methyltransferase, COMT) are intracellularly located, highly efficient uptake mechanisms exist to translocate the extracellular transmitter to the intracellular enzyme(s). Two such carrier-mediated uptake mechanisms are well known: a) the cocaine- and desipramine-sensitive, sodium- and chloride-dependent neuronal uptake (uptake$_1$) and b) the corticosterone- and 3-O-methyl-isoprenaline-sensitive, sodium- and chloride-independent extraneuronal uptake (uptake$_2$, Iversen, 1967). While uptake$_1$ translocates the amine to the intraneuronal MAO, uptake$_2$ transports the transmitter to both, extraneuronal COMT and extraneuronal MAO (Trendelenburg, 1984).

In such a metabolizing system, the rate of metabolism of noradrenaline is obviously dictated largely by the rate of inward transport, *i.e.*, by the kinetic constants of the carrier and by the extracellular amine concentration (So). In the following, an attempt is made to obtain an estimate of the activity of the intracellular enzymes that is independent of rates of inward transport. This is possible because of a characteristic relationship between "enzyme activity" and Si (intracellular amine concentration). For any given rate of inward transport, a highly active intracellular enzyme keeps Si low, while Si would be high if the intracellular enzyme had a low activity.

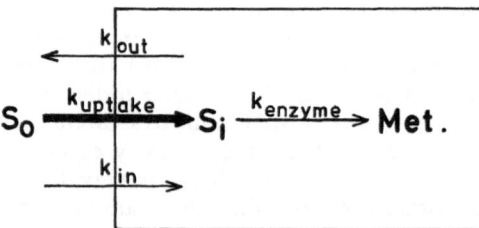

Fig. 1. Schematic representation of a "metabolizing system" exposed to such low an amine concentration (S_o) that neither uptake nor enzyme is saturated. There is inward transport of the substrate (characterized by $k_{uptake} = Vmax/Km$) as well as a diffusional influx (characterized by k_{in}). Once inside the "metabolizing system", the intracellular amine (S_i) either diffuses out (a process characterized by k_{out}) or is metabolized by the intracellular enzyme (characterized by $k_{enzyme} = Vmax/Km$) to yield the metabolite(s) *(Met)*

In the following, estimates of k_{enzyme} are obtained, *i.e.*, of that rate constant which characterizes the unsaturated enzyme (where $k_{enzyme} = Vmax/Km$). This rate constant (rather than Vmax) is of interest, since—under physiological conditions—transmitter-metabolizing enzymes appear to be exposed to transmitter concentrations well below Km; or in other words, they are not saturated normally.

k_{enzyme} of the extraneuronal and neuronal enzymes of the rat heart

Under steady-state conditions the rate of the appearance of the metabolite in the venous effluent of the perfused heart equals the rate of formation of the metabolite. Provided So is so low that the intracellular enzyme is not even partly saturated, we can write

$$v_{st-st} = Si \cdot k_{enzyme} \qquad (1)$$

where v_{st-st} = steady-state rate of metabolism, and $k_{enzyme} = Vmax/Km$. Thus, an estimate of k_{enzyme} (*i.e.*, of the activity of the intracellular enzyme) can be obtained from measurements of v_{st-st} and Si.

Hearts of reserpine-pretreated rats were perfused (Langendorff technique) with 50 nmol/l of various ^{3}H-catecholamines. Since tritium in position 8 (*i.e.*, on the alpha-carbon) is known to hinder the deamination of catecholamines (Trendelenburg *et al.*, 1983; Grohmann *et al.*, 1986), it should be emphasized that the ^{3}H-(—)-noradrenaline, ^{3}H-(—)-adrenaline and ^{3}H-dopamine used here was exclusively labelled in position 7 (*i.e.*, on the beta-carbon). The perfusion was long enough for the rates of appearance of the ^{3}H-metabolites in the venous effluent to reach steady state. In each experimental group only one uptake mechanism and one intracellular enzyme was intact—all other uptake mechanisms or enzymes were inhibited.

k_{comt} (extraneuronal system)

Neuronal uptake was inhibited by the presence of $30\,\mu$mol/l cocaine, MAO by pretreatment of the animals with 100 mg/kg pargyline. On perfusion with ^{3}H-catecholamines, only O-methylated ^{3}H-metabolites were detected (after their separation from the parent amine by the column chromatographic procedure of Graefe *et al.*, 1973). Table 1 presents v_{st-st}, Si and k_{comt} for the four ^{3}H-catecholamines. The activity of the enzyme declined in the order indicated in Table 1A, and it should be realized that a k_{comt} of 0.779 min^{-1} means that the intracellular COMT is able to O-methylate—per min—77.9% of the intracellular ^{3}H-(±)-isoprenaline.

Table 1. The rate constants describing the ^{3}H-catecholamine-metabolizing enzymes of the rat heart

^{3}H-catecholamine	n	$v_{st\text{-}st}$ (pmol·g^{-1}·min^{-1})	Si (pmol/g)	k_{enzyme} (min^{-1})
A. Extraneuronal COMT				
($\pm$)-isoprenaline	5	39.0 $\pm$ 1.8	50.0 $\pm$ 4.6	0.779 $\pm$ 0.043
(−)-adrenaline	5	42.1 $\pm$ 3.4	86.5 $\pm$ 9.0	0.487 $\pm$ 0.031
dopamine	5	16.5 $\pm$ 0.9	53.0 $\pm$ 3.7	0.312 $\pm$ 0.028
(−)-noradrenaline	5	23.5 $\pm$ 2.0	96.5 $\pm$ 2.8	0.244 $\pm$ 0.017
B. Extraneuronal MAO				
dopamine	4	14.1 $\pm$ 0.4	54.0 $\pm$ 2.6	0.261 $\pm$ 0.007
(−)-noradrenaline	4	21.3 $\pm$ 2.0	199.0 $\pm$ 29.6	0.107 $\pm$ 0.022
(−)-adrenaline	4	17.7 $\pm$ 1.4	337.0 $\pm$ 16.4	0.053 $\pm$ 0.006
C. Neuronal MAO				
(−)-noradrenaline	4	38.6 $\pm$ 3.1	110.5 $\pm$ 16.8	0.349 $\pm$ 0.038
dopamine	4	61.4 $\pm$ 6.1	196.0 $\pm$ 22.9	0.313 $\pm$ 0.039
(−)-adrenaline	4	5.1 $\pm$ 0.6	47.4 $\pm$ 3.7	0.107 $\pm$ 0.002

Rat hearts, perfused with 50 nmol/l ^{3}H-catecholamines until steady-state rates of metabolism were obtained. ^{3}H-catecholamine content of tissue was measured at the end of the experiment (and corrected for extracellular space). k_{enzyme} was calculated from $v_{st\text{-}st}$/Si (steady-state rate of metabolism/tissue content in steady-state). Vesicular uptake was inhibited in all experiments (pretreatment with reserpine); MAO was inhibited in A (by pretreatment with pargyline), COMT in B and C (by the presence of 10 μmol/l U-0521). Cocaine (10 μmol/l) was present in A and B (to inhibit neuronal uptake), 3-O-methylisoprenaline (100 μmol/l) in C (to inhibit extraneuronal uptake). Shown are means $\pm$ S.E. for the steady-state rate of metabolism ($v_{st\text{-}st}$), for the tissue content at the end of the experiment (Si) and for k_{enzyme} (= V max/Km), determined in n experiments.

k_{mao} (extraneuronal system)

Neuronal uptake was inhibited by the presence of 30 μmol/l cocaine, COMT by the presence of 10 μmol/l U-0521 (3, 4-dihydroxy-2-methyl propiophenone). For ^{3}H-(−)-noradrenaline and ^{3}H-(−)-adrenaline deamination resulted overwhelmingly in the production of ^{3}H-DOPEG (3, 4-dihydroxyphenylglycol), while ^{3}H-dopamine was deaminated mainly to the acid metabolite (^{3}H-DOPAC, 3, 4-dihydroxyphenylacetic acid). For the calculation of $v_{st\text{-}st}$ (Table 1) the minor deaminated metabolites were included. Table 1 B shows that—for all three catecholamines—$k_{mao\ extraneuronal}$ was substantially lower than $k_{comt\ extraneuronal}$.

k_{mao} (neuronal system)

Extraneuronal uptake was inhibited by the presence of 100 μmol/l 3-O-methylisoprenaline, while COMT was inhibited as described above. Again, ^{3}H-DOPEG and ^{3}H-DOPAC were the preferred deaminated metabolites, respectively. Again, v_{st-st} (Table 1 C) includes the deaminated metabolites of minor importance.

For all three ^{3}H-catecholamines the neuronal MAO activity exceeded that of the extraneuronal MAO. Moreover, it is of interest to note that ^{3}H-(−)-adrenaline was a poorer substrate of extraneuronal (Table 1 B) and neuronal (Table 1 C) MAO than was ^{3}H-(−)-noradrenaline.

A preliminary account of these findings was presented by Grohmann (1986).

The influence of k_{mao} on the axoplasmic ^{3}H-(−)-noradrenaline concentration

As stated above, a very highly active intracellular enzyme should keep the axoplasmic concentration of ^{3}H-(−)-noradrenaline very low. Moreover, it can be postulated that the inhibition of such a highly active intracellular enzyme should lead to very considerable rises of the axoplasmic ^{3}H-amine concentration. However, as indicated by equation (1), a pronounced decline in k_{mao} (i.e., pronounced inhibition of MAO) should have a relatively small effect on v_{st-st}, if the pronounced decline in k_{mao} is accompanied by a substantial increase in Si. Experiments were carried out to test this working hypothesis (Cassis et al., 1986).

Lengthwise halved vasa deferentia of reserpine-pretreated rats were pre-incubated (for 30 min) with 0, 10, 20 or 40 nmol/l pargyline and then washed out (for 30 min). Subsequently the tissues were exposed to 10 nmol/l ^{3}H-(−)-noradrenaline (+ 100 μmol/l ^{14}C-sorbitol) for 300 min (with transfer to a new solution every 30 min). Each incubation medium and the radioactivity extracted from the tissue were subjected to column chromatography. Of the deaminated ^{3}H-metabolites, ^{3}H-DOMA accounted for only 0.1% of total ^{3}H-metabolite formation; it is disregarded in the following.

Fig. 2 A shows the cumulative formation of ^{3}H-DOPEG. When MAO was intact, there was no initial delay of the deamination of ^{3}H-(−)-noradrenaline, and the ^{3}H-amine was deaminated at a steady rate throughout the experiment. Note that the slope of the straight line indicates the rate of steady-state deamination. After inhibition of MAO by 10, 20 or 40 μmol/l pargyline, clear evidence of enzyme inhibition was obtained after the first 30 min period of incubation: there was a concentration-dependent decline of the formation of ^{3}H-DOPEG. However, subsequently the rates of deamination accelerated for a considerable time, in order to reach steady state

U. Trendelenburg *et al.*

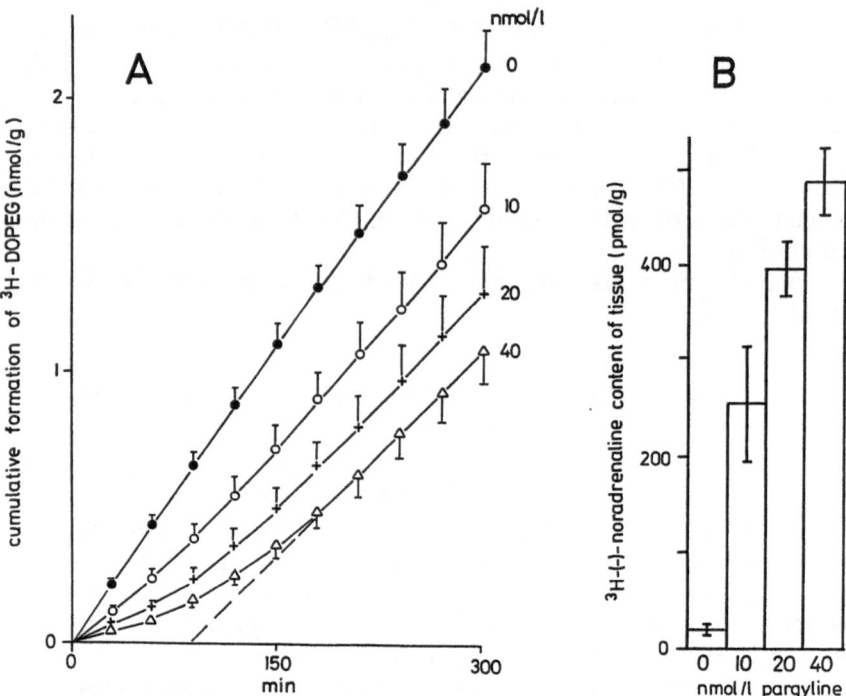

Fig. 2. The effect of inhibition of MAO on the cumulative neuronal deamination of
^{3}H-(−)-noradrenaline **(A)** and on the ^{3}H-(−)-noradrenaline content of the neuronal
tissue **(B)**. Rat vas deferens after inhibition of vesicular uptake and COMT, exposed
to 10 nmol/l ^{3}H-(−)-noradrenaline for 300 min (= 10 successive periods of 30 min
each). Tissues were preincubated with 0, 10, 20 or 40 nmol/l pargyline (for 30 min)
and then washed out (for 30 min). Tissue content was measured at the end of the experiment and corrected for the extracellular distribution of the ^{3}H-amine. n = 4 to 8 for
each mean value. **A** Ordinate: cumulative formation of ^{3}H-DOPEG (recovered from
medium and tissue, in nmol/g). Abscissa: min of exposure to 10 nmol/l ^{3}H-(−)-nor-
adrenaline. **B** Height of columns indicates ^{3}H-(−)-noradrenaline content of tissue at
end of experiment. The broken line in A illustrates that it took the rate of deamination more than 150 min to reach steady state. Reproduced from Cassis *et al.* (1986),
with the publishers' permisson

after a considerably delay. Parallel experiments with homogenates of
similarly treated vasa deferentia indicated that preincubation with
10 nmol/l pargyline caused 83 to 89% inhibition of the deamination
of either ^{3}H-(−)-noradrenaline or ^{3}H-5-hydroxytryptamine, while the
corresponding values were 93 and 96% after preincubation with
20 nmol/l pargyline.

How can one reconcile an about 98 or 99% inhibition of MAO

(with 40 nmol/l pargyline) with just a halving of the steady-state rate of deamination (see slopes of curves in Fig. 2 A)? The answer to this question is found in the very pronounced rise in Si, when MAO is inhibited. As shown in Fig. 2 B, inhibition of MAO resulted in very pronounced increases of the ^{3}H-(−)-noradrenaline distributed into the nerve endings. This increase explains why there was no dramatic decline in v_{st-st}, it also explains the pronounced delay in the attainment of steady state (see Fig. 2 A, the time needed for the three lower curves to become straight lines): considerable time is needed for the build up of the very high axoplasmic concentration of ^{3}H-(−)-noradrenaline.

From the results of Figs. 2 A and B it is possible to calculate $k_{mao\ neuronal}$ (for ^{3}H-[−]-noradrenaline), which amounted to 0.47 min^{-1}; this value does not differ much from that determined for the adrenergic nerve endings of the rat heart (Table 1 C). However, it should be realized that all calculations of $k_{mao\ neuronal}$ (rat heart or vas deferens) were based on the assumption that we deal with a "one compartment-system" (as indicated by Fig. 1). This assumption is wrong. Although the pretreatment with reserpine prevented any transport of the ^{3}H-catecholamines into the storage vesicles of the adrenergic nerve endings, ^{3}H-(−)-noradrenaline (as an example) is able to diffuse through the vesicular membrane, although at a low rate. Since the pH inside the storage vesicles is very low (pH 5.5), ^{3}H-(−)-noradrenaline is trapped inside the vesicles (because it is nearly fully dissociated at the low pH, hence hardly able to diffuse out of them). As "vesicular" ^{3}H-(−)-noradrenaline is protected from mitochondrial MAO, any vesicular trapping of the amine leads to an underestimate of $k_{mao\ neuronal}$. In separate and yet incomplete experiments estimates of the relative distribution of ^{3}H-(−)-noradrenaline into the axoplasmic and vesicular compartments were obtained: about 25% of the amine is in the axoplasm, 75% in the storage vesicles. Hence, if the two estimates of $k_{mao\ neuronal}$ (rat heart and rat vas deferens) gave a value of about 0.4 min^{-1}, the correct value is about 4 times higher, or 1.6 min^{-1}. Neuronal MAO then is of such high activity that it can deaminate, per min, about 160% of the axoplasmic ^{3}H-(−)-noradrenaline.

These results illustrate the two sides of the same coin: on the one hand, there is an impressive activity of MAO inside the varicosities of adrenergic nerve endings; on the other hand, this impressive activity is responsible for keeping the axoplasmic noradrenaline concentration at a very low level. However, as soon as MAO is even partly inhibited, a very substantial increase in the axoplasmic amine concentration results.

Is neuronal MAO saturated (inhibited) by the inward transport of substrates of the enzyme?

Whenever an experimental procedure results in a decrease of the rate of deamination of ^{3}H-(−)-noradrenaline, experimenters have been apt to conclude that MAO was inhibited. For instance, Leitz and Stefano (1971) observed that tyramine reduced the rate of deamination of ^{3}H-noradrenaline, and they proposed that a) the neuronal uptake of tyramine is able to inhibit (*i.e.*, to saturate) neuronal MAO and that b) this inhibition of neuronal MAO is somehow responsible for the well-known noradrenaline-releasing effect of tyramine.

With this in mind, it is important to emphasize that the rate of neuronal deamination of ^{3}H-(−)-noradrenaline can decline for a very different reason. Stute and Trendelenburg (1984) loaded rat vasa deferentia with ^{3}H-(−)-noradrenaline and then washed the tissues with amine-free solution. It should be noted that the results described in this section were all obtained with tissues from rats that had not received any pretreatment with reserpine; hence, vesicular uptake was left intact. From the 100th min of wash-out onwards, one then obtains a "neuronal efflux of radioactivity" that is characterized by a) an FRL (fractional rate of loss = rate of efflux divided by tissue content determined at the same time) that is constant with time, and b) by a rate of efflux of ^{3}H-DOPEG that is substantially higher than that of ^{3}H-(−)-noradrenaline. Because the free neuronal carrier is found on that side of the axonal membrane on which the sodium concentration is high (*i.e.*, normally on the outside; Sammet and Graefe, 1979), any increase in the internal sodium concentration brings free carrier to the inside of the axonal membrane, where it can then sustain an outward transport of the preloaded ^{3}H-(−)-noradrenaline. However, such experiments must be carried out in the absence of extracellular calcium, so as to prevent any exocytotic release of ^{3}H-(−)-noradrenaline. Stute and Trendelenburg (1984) increased the inside sodium concentration either by inhibition of the Na^+-K^+-ATPase (by reduction of the extracellular potassium concentration or by ouabain) or by veratridine (which opens the fast sodium channels). In each case, the ensuing outward transport of ^{3}H-(−)-noradrenaline was accompanied by a decline of the efflux of ^{3}H-DOPEG. As it is most unlikely that ouabain or veratridine or sodium can inhibit neuronal MAO, it is permitted to conclude that any pronounced outward transport of ^{3}H-(−)-noradrenaline lowers the axoplasmic ^{3}H-amine concentration, *i.e.*, the substrate concentration around the mitochondrial enzyme. Moreover, it can be stated that outward transport of ^{3}H-(−)-noradrenaline goes hand in hand with an increase in the ratio NA/DOPEG (FRL for ^{3}H-[−]-noradrenaline/FRL for ^{3}H-DOPEG).

In a further series of experiments, the outward transport of ^{3}H-(−)-noradrenaline was induced by twelve different substrates of the neuronal carrier ([+]-amphetamine, amezinium, [−]-metaraminol, tyramine, dopamine, [−]-noradrenaline, debrisoquin, [±]-phenylephrine, guanethidine, bretylium, bethanidine and 5-hydroxytryptamine—in order of increasing Km for uptake$_1$). Whenever outward transport was induced, the ratio NA/DOPEG increased; moreover, the magnitude of this increase was correlated with the magnitude of the ^{3}H-(−)-noradrenaline-releasing effect of various concentrations of these amines. Interestingly enough, the twelve agents belonged to two groups. Five of them were substrates of MAO (tyramine, dopamine, [−]-noradrenaline, [±]-phenylephrine, 5-hydroxytryptamine), while two were neither substrates nor inhibitors of MAO ([−]-metaraminol, guanethidine). For these seven compounds virtually identical increases in the ratio NA/DOPEG were obtained (for equal rates of ^{3}H-(−)-noradrenaline release). The other five compounds ([+]-amphetamine, amezinium, debrisoquin, bretylium and bethanidine) are all known to inhibit MAO. They all caused increases in the ratio NA/DOPEG that were larger than for the first mentioned group. Obviously, these compounds increase the ratio NA/DOPEG not only by the induction of an outward transport of ^{3}H-(−)-noradrenaline, but also by inhibition of intraneuronal MAO.

From this relationship we are permitted to conclude that intraneuronal MAO is not saturated (or inhibited) by tyramine, even when this compound is transported into the nerve ending at rates close to its Vmax. Tyramine was administered in a concentration of 1000 μmol/l; since its Km for uptake$_1$ was 1.4 μmol/l, its rate of uptake (at 1000 μmol/l) amounts to 99.9% of Vmax. As in the preceding sections of this report, it has to be stressed that the intraneuronal MAO activity is very high—obviously high enough to prevent the build up of a saturating tyramine concentration in the axoplasm, when the amine is transported into the nerve ending at its Vmax.

A preliminary account of these experiments was presented by Langeloh (1986).

Conclusions

The results presented here serve to illustrate some of the complexities that arise when a saturable uptake mechanims is arranged in series with a saturable (intracellular) enzyme. As far as intraneuronal MAO is concerned, three different series of experiments gave evidence for a very high intraneuronal MAO activity. In the extraneuronal metabolizing system, on the other hand, COMT is clearly the enzyme that is considerably more important than is MAO. One consequence of this very high intraneuronal MAO activity is the ability of this enzyme to keep the axoplasmic noradrenaline at a very low level. If one considers the fact that the exocytotic release of noradrenaline (induced by action potentials which reach the varicosities with a high frequency) can easily lead to concentrations of noradrenaline in the extracellular space that can largely saturate uptake$_1$, a very high intraneuronal MAO activity can ensure that—inspite of nearly maximal rates of uptake of noradrenaline—the axoplasmic concentration stays low. Hence, there is no danger of any leakage of noradrenaline from the nerve endings between two impulses—a leakage that would prevent the sympathetic innervation from controlling the function of sympathetically innervated organs.

Note added in proof

After pretreatment of the rats with reserpine and after loading the nerve endings with ^{3}H-(−)-noradrenaline, the "neuronal amine" (= Si) distributes (under steady-state conditions) into axoplasm and storage vesicles in the ratio $1:1.2$. Hence, the k_{mao}-values in Table 1 C have to be multiplied by a factor of 2.2 (and not by a factor of 4, as stated in the text). Thus, $k_{mao\ neuronal}$ amounts to values between 0.77 and 0.24 min^{-1} and exceeds $k_{mao\ extraneuronal}$ by factors of 2.9 to 6.4.

References

Cassis L, Ludwig J, Grohmann M, Trendelenburg U (1986) The effect of partial inhibition of monoamine oxidase on the steady-state rate of deamination of ^{3}H-catecholamines in two metabolizing systems. Naunyn-Schmiedeberg's Arch Pharmacol 333: 253—261

Graefe K-H, Stefano FJE, Langer SZ (1973) Preferential metabolism of (−)-^{3}H-norepinephrine through the deaminated glycol in the rat vas deferens. Biochem Pharmacol 22: 1147—1160

Grohmann M (1986) Neuronal and extraneuronal metabolism of different catecholamines (CA) in the perfused rat heart. Naunyn-Schmiedeberg's Arch Pharmacol 332: R 74

Grohmann M, Henseling M, Cassis L, Trendelenburg U (1986) Errors introduced by a tritium label in position 8 of catecholamines. Naunyn-Schmiedeberg's Arch Pharmacol 332: 34—42

Iversen LL (1967) The uptake and storage of noradrenaline in sympathetic nerves. Cambridge University Press, Cambridge

Langeloh A (1986) The mechanism of action of indirectly acting sympatho-mimetic amines. Naunyn-Schmiedeberg's Arch Pharmacol 332: R 75

Leitz FH, Stefano FJE (1971) The effect of tyramine, amphetamine and meta-raminol on the metabolic disposition of ^{3}H-norepinephrine released from the adrenergic neuron. J Pharmacol Exp Ther 178: 464–473

Sammet S, Graefe K-H (1979) Kinetic analysis of the interaction between noradrenaline and Na$^+$ in neuronal uptake: kinetic evidence for co-transport. Naunyn-Schmiedeberg's Arch Pharmacol 309: 99–107

Stute N, Trendelenburg U (1984) The outward transport of axoplasmic nor-adrenaline induced by a rise of the sodium concentration in the adrener-gic nerve endings of the rat vas deferens. Naunyn-Schmiedeberg's Arch Pharmacol 327: 124–132

Trendelenburg U (1984) Metabolizing systems. In: Fleming WW, Langer SZ, Graefe K-H, Weiner N (eds) Neuronal and extraneuronal events in autonomic pharmacology. Raven Press, New York, pp 93–109

Trendelenburg U, Stefano FJE, Grohmann M (1983) The isotope effect of tritium in ^{3}H-noradrenaline. Naunyn-Schmiedeberg's Arch Pharmacol 323: 128–140

Authors' address: Dr. U. Trendelenburg, Pharmakologisches Institut, Universität Würzburg, Versbacher Strasse 9, D-8700 Würzburg, Federal Republic of Germany.

J Neural Transm (1987) [Suppl] 23: 103–119

Overview of the present state of MAO inhibitors

Margherita Strolin Benedetti and P. Dostert

Laboratoires Fournier, Centre de Recherches de Daix, Fontaine-les-Dijon, France

Summary

In this paper an overview of the present state of monoamine oxidase inhibitors (MAOIs) is presented. The irreversible inhibitors are firstly considered. They have been divided into four chemical types: substituted hydrazine, cyclopropylamine, propargylamine and allylamine derivatives. Moreover, a tetrahydropyridine derivative (MPTP), recently described as an irreversible inhibitor of MAO-B, has been included among the irreversible MAOIs.

The reversible inhibitors such as tetrahydro-β-carbolines and salsolinol, phenylalkylamines: amphetamine, amiflamine and 2, 3-dichloro-α-methylbenzylamine.

Among the short acting or reversible inhibitors the 4-(2-benzofuranyl) piperidine series and the morpholinoethylamino derivatives are discussed.

Finally the oxazolidinone series is presented separately, as in this series reversible or irreversible inhibitors of the A or B form of MAO have been obtained.

The purpose of this paper is to present an overview of the present state of monoamine oxidase inhibitors (MAOIs).

The generally used definition of the two forms of MAO is that MAO-A is inhibited by low (around nanomolar) concentrations of clorgyline whereas MAO-B is not inhibited until micromolar concentrations of this inhibitor are used. The reverse is true when l-deprenyl is used to inhibit MAO activity. Although substrates of the A and B forms of MAO do not show absolute specificity, β-phenylethylamine (PEA) may be considered as a preferential B substrate and 5-hydroxytryptamine (5-HT) as a preferential A substrate. According to Russel *et al.*, (1979) the majority of the PEA oxidizing activity of MAO (MAO-B) seems to be situated on the outer surface and the majority of the 5-HT oxidizing activity (MAO-A) on the inner surface of the mitochondrial outer membrane, but Buckman *et al.* (1984) have suggested that this may not in fact be the case.

There was no unequivocal evidence that the oxidases could be physically separated as catalytically active species which also retained native enzymatic properties until the recent report of Pearce and Roth, 1984. These authors demonstrated that the oxidases are separable by simple chromatographic techniques. This and other recent studies, particularly those where monoclonal antibodies specific for the two forms of the enzyme have been used (Denney *et al.*, 1983; Westlund *et al.*, 1985), have provided sufficient evidence to strongly suggest that the two major enzyme subtypes may be different proteins.

The reaction catalyzed by MAO can be described as a two step process:

1. $R-CH_2-NH_2+O_2 \rightarrow [R-CH=NH]+H_2O_2$ (enzymatic)
2. $[R-CH=NH]+H_2O \rightarrow R-CHO+NH_3$ (non enzymatic)

As shown by Belleau and Moran (1963) with tyramine and confirmed by Yu *et al.* (1986) with dopamine, removal by MAO of an α-hydrogen from a substrate is stereospecific and involves the pro-R-hydrogen.

Irreversible inhibitors

Four chemical types of suicide substrates or suicide enzyme inactivators have been described for MAO, namely substituted hydrazine, cyclopropylamine, propargylamine and allylamine derivatives. Moreover, a tetrahydropyridine derivative, MPTP, has been recently described as an irreversible inhibitor of MAO-B.

Hydrazines

The hydrazines were the first suicide enzyme inactivators to be recognized. They are non selective inhibitors. Iproniazid by hydrolysis gives isopropylhydrazine which through FAD is transformed into isopropyl diazene, which inhibits the enzyme (Alston, 1981). Phenelzine is oxidized either to the inert hydrazone or the reactive diazene (Alston, 1981). The formation of hydrazone might explain the presence *"in vivo"* of phenylacetic acid as a metabolite of phenelzine. Phenylhydrazine, after oxidation to the corresponding dia-zene, partially attacks the nucleophilic C(4 a) position of the flavin cofactor of MAO in a reaction which involves loss of nitrogen and partially reacts with the enzyme protein, probably at a thiol group (Kenney *et al.*, 1979; Singer and Salach, 1981).

Cyclopropylamines

MAO inactivation by tranylcypromine, another non selective irreversible inhibitor, seems to involve oxidation to the imine which either as such or after hydrolysis to the ketone reacts with an -SH group at the substrate site of MAO to form a thioaminoketal or thio-hemiketal (Paech *et al.*, 1980). However, according to Silverman (1983), cyclopropyl ring opening of tranylcypromine occurs and the resulting adduct with an SH group liberates cinnamaldehyde on dis-sociation. For N-cyclopropyl-N-arylalkylamines, it is postulated that the cyclopropyl carbon attached to the nitrogen is oxidised, yielding the highly reactive cyclopropanone imine which reacts either with N-5 of the reduced form of flavin or with a nucleophile site of the protein (Silverman and Hoffman, 1980). In this group, selective in-hibitors have been reported, such as Lilly 51641 and Lilly 54761 which are respectively MAO-A and MAO-B inhibitors (Fowler and Ross, 1984). In the case of l-phenylcyclopropylamine, eight molecu-les are needed to irreversibly inactivate each MAO enzyme molecule. One molecule reacts with the FAD cofactor at the N-5 position to give an irreversible adduct; the other seven form a reversible adduct with a cysteine residue (Silverman and Zieske, 1986).

Acetylenic inhibitors

MAO inactivation by these compounds consists of two phases: an instantaneous reversible one, which is competitive with the sub-strate, and an irreversible phase which involves oxidation of the amine function to imine by flavin, followed by formation of an ir-reversible adduct with the isoalloxazine moiety of FAD (Singer, 1979). Numerous analogues of deprenyl have been synthesized by Magyar *et al.* (1979). Some of them displayed an inhibitory effect on MAO-B superior to that of deprenyl, whereas others were found to be potent MAO-A inhibitors.

Allylamine derivatives

If oxidative deamination of a substrate occurs through imine and aldehyde formation, then the corresponding β-halomethylene analogues would be predicted to act as suicide substrates, leading to irreversible inactivation (Palfreyman *et al.*, 1986). This approach came from the work of Rando and Eigner (1977) who used chloro- and bromo-allylamines to weakly but irreversibly inhibit MAO. The mechanism by which cis-3-chloroallylamine inhibits MAO seems to be identical with that of β, γ-acetylenic amines. Using the selectivity of substituted phenylethylamine substrates, Palfreyman *et al.* (1986) could obtain selective inhibitors. Using aminoacid bioprecursors of their compounds with an inhibitor of extracerebral L-amino acid decarboxylase, they obtained potent central inhibitors of MAO. Their MDL 72392 nevertheless did show modest selectivity for the A-form of the enzyme. As suggested by Kalir *et al.* (1981) and by Knoll *et al.* (1978), selectivity of MAOIs could be related in part to the distance between the aromatic ring and the nitrogen atom and the type A selectivity of clorgyline attributed to the 4-atom linkage between the dichlorobenzene ring and the nitrogen atom. This led McDonald *et al.* (1986) to prepare a structurally related fluoroallylamine (MDL 72638) in an attempt to improve the modest A-selectivity seen with MDL 72392. In contrast to what was expected, MDL 72638 was found to be a potent and very selective inhibitor of the B-form of MAO.

1-methyl-4-phenyl-1, 2 3, 6-tetrahydropyridine (MPTP)

MPTP was studied *in vitro* as an inhibitor of type A or B MAO of rat brain using 5-HT or PEA as substrate, respectively (Fuller and Hemrick-Luecke, 1985). MPTP was less potent inhibitor of MAO-B than of MAO-A. The inhibition of MAO-B showed noncompetitive kinetics, was not fully reversible by dialysis and was time dependent. In contrast MPTP was a competitive (Ki $=$ 9 μM) and reversible inhibitor of MAO-A. MPTP is metabolized by MAO-B of rat brain and liver mitochondria (Chiba *et al.*, 1984; Boucher *et al.*, in press).

Reversible inhibitors

Harmala alkaloids

Harmine and harmaline are selective, potent and reversible inhibitors of MAO-A (Fowler and Ross, 1984). According to these properties, the tritiated derivative of harmaline was considered as an appropriate ligand for exploring MAO-A in tissues (Nelson *et al.*, 1979; Kan and Strolin Benedetti, 1981).

Endogenous inhibitors of MAO

Tetrahydro-β-carbolines (THBCs) are structurally similar to the harmala alkaloids. Among them, some with differing degrees of MAO inhibitory potency have been identified in various tissues and fluids from animals and man and can be considered as potential endogenous MAOIs. These compounds can be obtained by the reaction of an aldehyde or an α-ketoacid with an indole ethylamine. THBCs competitively inhibit the oxidative deamination of 5-HT at μM concentrations. In contrast, effective inhibition of PEA deamination was observed only at or near millimolar concentrations of these compounds (Meller *et al.*, 1977).

A remarkable structural similarity exists between the parkinsonian neurotoxin, MPTP, and 2-methyl THBC. In a recent paper, Collins and Neafsey (1985) suggest that N-methylated β-carboline species, possibly accumulating during stress and aging, could well be causative factors in Parkinsonism.

Salsolinol (Sal) is an endogenous compound which might be formed by condensation of dopamine with acetaldehyde and/or pyruvic acid. According to Meyerson *et al.* (1976), Sal inhibits the oxidative deamination of 5-HT in a competitive manner with an apparent K_i value of $110\,\mu M$, whereas it noncompetitively inhibits benzylamine oxidation with an apparent K_1 value of $52\,mM$. Although Sal has a very weak inhibitory activity, preliminary data (Strolin Benedetti and Boucher, unpublished results) show that this activity is largely due to the enantiomer with the R configuration. As the asymmetrical centre in Sal is at the α carbon of the benzylamine moiety of the molecule, it would be interesting to know whether the proton of benzylamine abstracted by MAO-B corresponds to the pro-S or to the pro-5 hydrogen (Fig. 1).

Phenylalkylamines

In the phenylalkylamine type substrates the substitution of an hydrogen in α position to the nitrogen atom by a methyl group, has resulted in the formation of MAOIs. The selectivity of the α-methyl

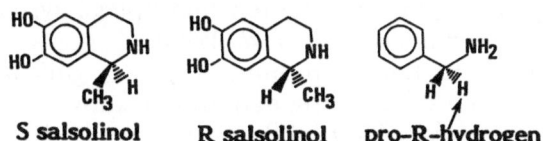

S salsolinol   R salsolinol   pro–R–hydrogen

Fig. 1. Structure of benzylamine and of enantiomers of salsolinol

amine derivatives for one form of MAO can be improved by resolution of the two optical isomers.

Phenylethylamines

Introduction of a methyl group in the α position in PEA, a preferential B substrate, inhibits its oxidative deamination by MAO and produces amphetamine, a preferential inhibitor of the A-form. The ratio $\frac{Ki(MAO-B)}{Ki(MAO-A)}$ for the S enantiomer is 38 and for the R enantiomer 8.5 (Strolin Benedetti and Dostert, 1985). When the methyl group occupies the position of the non-abstracted hydrogen in the substrate, a greater affinity of the inhibitor for the active site of MAO-A is obtained. However, the difference in affinity between the R and S enantiomers of amphetamine for the A- or B-form of MAO is small; for MAO-A: $\frac{Ki(S)}{Ki(R)} = 3.5$ and for MAO-B: $\frac{Ki(R)}{Ki(S)} = 1.2$.

Saturable and stereospecific binding sites for $(+)^3$H-amphetamine were recently demonstrated by two different teams (Paul *et al.*, 1982; Lesage *et al.*, 1984). As ^{3}H-harmaline was used to label the MAO-A site and as this compound, as well as other MAO-A selective inhibitors, is a good displacer of ^{3}H-amphetamine, MAO-A might be a component of the complex mixture of binding sites recognized by ^{3}H-amphetamine. Arguments presented by Lesage *et al.* (1985) against this possibility would need further confirmation. S(+)FLA 336 (amiflamine) is also a more potent and selective MAO-A inhibitor than its R enantiomer. Like amphetamine, the more potent and selective enantiomer of FLA 336 has the α-hydrogen in the same steric configuration as the pro-R-hydrogen of the phenylethylamine substrates of MAO (Strolin Benedetti and Dostert, 1985).

With respect to the noradrenaline and serotonin hypothesis of affective disorders, amiflamine was found in the rat brain to be a more potent MAOI in serotonergic neurones than in noradrenergic and dopaminergic neurones, whereas N, N-didesmethylamiflamine [FLA 668(+)] inhibited preferentially MAO in noradrenergic neurones. The plausible explanation is that these MAOIs are transported selectively by the 5-HT or noradrenaline uptake system into the aminergic nerve terminals (Ask *et al.*, 1986).

Coming back to tranylcypromine (T), a "rigid" phenylethylamine analogue in a trans configuration, it is known that (+)T (1 S, 2 R) is a more powerful MAO inhibitor than (−)T (1 R, 2 S), both of A- and B-forms of MAO (Riley and Brier, 1972), although preincuba-

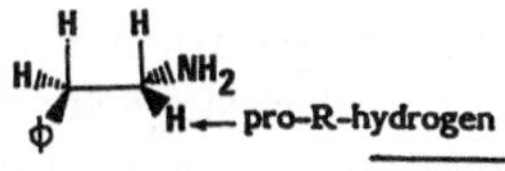

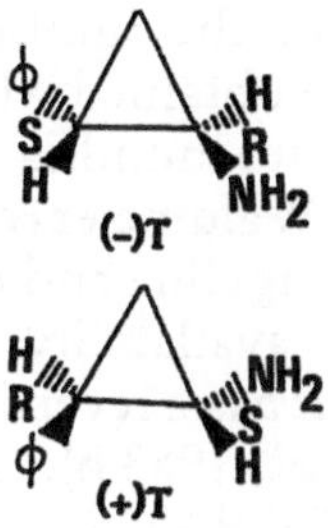

| | IC_{50} (M) | | | |
| | t = 0 | | t = 60 min. | |
	MAO-A	MAO-B	MAO-A	MAO-B
(−)T	$8\ 10^{-5}$	$3.5\ 10^{-5}$	$7\ 10^{-7}$	$6\ 10^{-7}$
(+)T	$2.5\ 10^{-6}$	$3.5\ 10^{-6}$	10^{-7}	$4\ 10^{-8}$
$\dfrac{(-)T}{(+)T}$	32	10	7	15

Fig. 2. MAO inhibition by the enantiomers of tranylcypromine

tion diminishes the difference of their inhibitory potency towards MAO-A (Strolin Benedetti et al., unpublished results). Here again, the pro-R-hydrogen of the substrate removed by MAO corresponds to that present in the S carbon of the (+)T (Fig. 2).

Benzylamines

Fuller and Hemrick (1978) studied the inhibition of MAO by the two enantiomers of 2, 3-dichloro-α-methylbenzylamine. The (+)isomer inhibited the oxidation of 5-HT by rat brain MAO more effectively than it inhibited the oxidation of PEA. In contrast, the (−)isomer inhibited PEA oxidation more than it inhibited 5-HT oxidation. Unfortunately the absolute configuration of the two enantiomers is not given.

Short acting or reversible inhibitors

4-(2-benzofuranyl) piperidine series

CGP 11305 A (brofaremine) inhibits 5-HT deamination in the brain and liver of pretreated rats with ED_{50} values of 1 and 0.7 mg/kg p.o., respectively. PEA deamination is only marginally affected up to doses of 100 mg/kg p.o. The duration of action of brofaremine is less than 48 hours and its effect does not cumulate after repeated oral administration. Kinetic studies in mitochondrial preparations have shown that, with both 5-HT and PEA as substrates, the inhibition is of the competitive type when the enzyme preparation and the inhibitor are not preincubated prior to assay. This suggests a reversible interaction with the enzyme. However, in *ex vivo* studies the inhibitory activity of brofaremine was not affected by dilution or dialysis of homogenates from pretreated animals and the inhibitor could not be displaced by 5-HT. Similar results were obtained when brofaremine was preincubated *in vitro* with mitochondria or homogenates, indicating an irreversible interaction (Waldmeier *et al.*, 1983 a). However, protection experiments against clorgyline and displacement of brofaremine by increasing the availability of endogenous substrates suggest that brofaremine *in vivo* is a reversible and competitive MAO-A inhibitor (Waldmeier *et al.*, 1983 b).

Brofaremine is a weak 5-HT uptake blocker. CGP 6085 A is a potent 5-HT uptake blocker with weak MAO-A inhibitory properties while CGP 4718 A possesses both MAO-A and 5-HT uptake-inhibiting properties in the same dose range (Ciba-Geigy, 1986).

Morpholinoethylamino derivatives

Moclobemide, despite its lack of activity *in vitro,* strongly inhibits MAO-A *ex vivo* (ED_{50} = 0.9 mg/kg i.p.). Minaprine behaves *in vitro* as a very weak and non specific MAOI. However, *ex vivo* this drug preferentially inhibits MAO-A, although with a mild potency (ED_{50} = 12.8 mg/kg i.p.) and with a short duration of action. Both *in vitro* and *ex vivo*, SR 95191, a minaprine analogue, is more potent and selective than minaprine towards MAO-A (ED_{50} = 7.5 mg/kg i.p.). The discrepancy between the *in vitro* and *ex vivo* results for moclobemide and minaprine suggest that one (or several) metabolites may be responsible for their MAO-A inhibitory effect (Bizière *et al.*, 1985).

MAO inhibition in rat liver and brain by the short-acting MAO-AIs moclobemide and brofaremine, administered p.o. at roughly equieffective doses 2 hours before decapitation, was investigated by Keller *et al.* (1986) for its reversibility under various conditions. MAO-A activity in liver and brain homogenates, inhibited by moclobemide (300 μmoles/kg) to approximately 15% of control, partially or completely returns to control value upon dialysis at 37 °C. Under these conditions, recovery of enzyme activity after brofaremine (30 μmoles/kg) is minimal.

Two very interesting new MAO-B inhibitors have been described recently; RO 16-6491 is chemically related to, and perhaps produced as a metabolite from moclobemide. A slight modification of RO 16-6491, has led to RO 19-6327, a more potent and selective MAO-B

inhibitor. RO 16-6491 inhibits very potently and selectively the deamination of PEA in rat brain homogenates as well as in human brain
cortex. In *ex vivo* experiments the compound is also highly selective.
The time course of MAO-B inhibition in brain and liver of rats administered RO 16-6491 [10 μmoles (2.4 mg)/kg p.o.] shows an inhibition of PEA deamination of about 80% between 30 min and 4 hours,
whereas after 16 hours MAO-B activity was fully recovered (Da Prada
et al., 1986). Cesura *et al.*, (1985) have shown that RO 16-6491 being a
reversible inhibitor of MAO-B, it can be used in its tritiated form as a
suitable ligand for exploring MAO-B.

As reported by Da Prada *et al.* (in press), Ro 19-6327 is the most
potent and selective short-lasting MAO-B inhibitor so far encounter-
ed.

Oxazolidinone series

The basic structure represented in Fig. 3 enabled reversible or
irreversible inhibitors of the A or B form of MAO to be obtained.
Some of them are devoid of the amine function, which generally
characterizes the MAOIs. A first study was carried out with R = OH,
A = N and B = O, by varying the position of R_1. Position 2 was not
convenient whereas positions 3 and 4 are practically equivalent for
MAO-A inhibition (Dostert and Strolin. Benedetti, 1986).

Toloxatone (R_1 = 3-CH$_3$) was found to be a reversible and competitive inhibitor of MAO-A. The Ki values of toloxatone and of its
enantiomers for MAO-A and B are presented in Fig. 4. These values
show that inhibition produced by the R enantiomer is stronger and
more selective for MAO-A than that produced by the S enantiomer.
The R enantiomer of toloxatone has the same absolute configuration
as R(−)NA. Toloxatone has been found to be an effective antidepressant and has now been marketed for this indication under the trade
name Humoryl* in France.

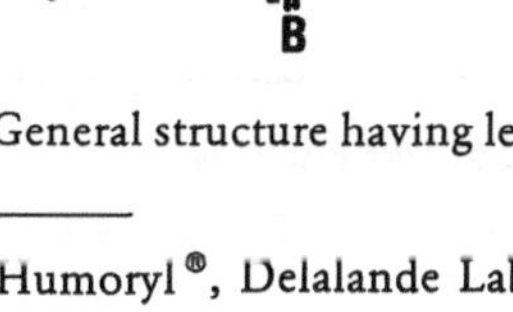

A   = CH, N
B   = O, S
R   = OH, OCH$_3$, N$<$
R_1 = alkyl, substituted benzyloxy ...

Fig. 3. General structure having led to reversible or irreversible MAOIs of the A or B
forms

* Humoryl®, Delalande Laboratories, France.

		Ki (μM)		Ki (MAO-B)
		MAO-A	MAO-B	Ki (MAO-A)
Toloxatone	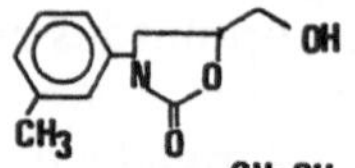	1.8	44	24
R enantiomer	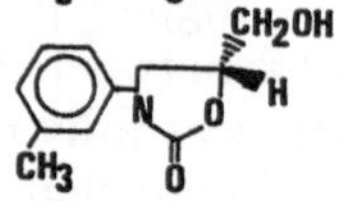	0.8	38.9	48
S enantiomer	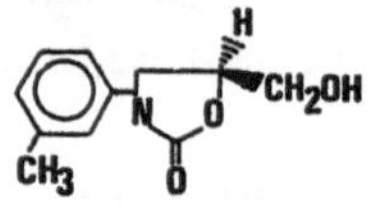	12.5	78.6	6

Fig. 4. Ki values of toloxatone and its enantiomers for MAO-A and -B

In a further study a series of benzyloxy derivatives (Fig. 3, $R_1 =$ $O - CH_2 - C_6H_5$) has been prepared by varying the substituent in the phenyl of the benzyloxy group. Again substitution in position 2 was not convenient whereas substitution in position 3 by a CN or by a chlorine atom produced potent MAO-A or MAO-B inhibitors, respectively. R enantiomers have the same absolute configuration as R(−)NA and are more potent inhibitors of MAO-A than their corresponding S enantiomers. The more potent and selective MAO-A inhibitor has the R configuration, whereas in the case of MAO-B inhibitors the more potent has the R configuration and the more selective the S configuration. MAO-A is more sensitive to the stereo-isomerism of these alcohol derivatives (β position) than MAO-B (Strolin Benedetti, 1984) (Fig. 5).

	Ki (μM)		Ki (MAO-B)	Ki (MAO-A)
	MAO-A	MAO-B	Ki (MAO-A)	Ki (MAO-B)
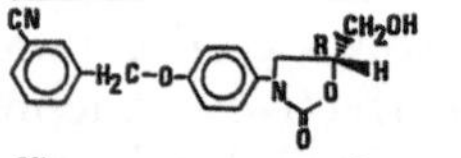	0.006	0.36	60	
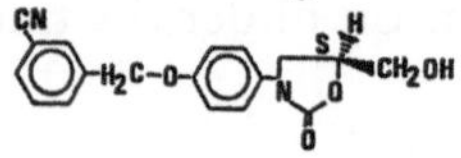	0.14	0.93	7	
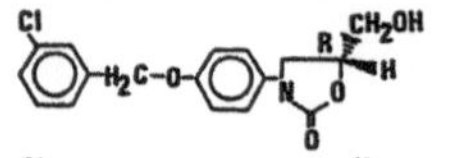	0.022	0.013		1.7
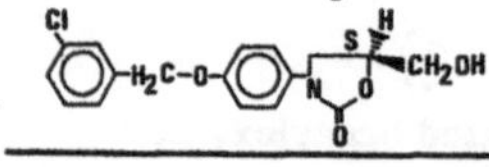	0.66	0.027		24

No preincubation of the inhibitor with the enzyme

Fig. 5. Ki values for MAO-A and -B of the enantiomers of two 5-hydroxymethyl derivatives in oxazolidinone series

| | Ki (μM) | | Ki (MAO-B) | Ki (MAO-A) |
	MAO-A	MAO-B	Ki (MAO-A)	Ki (MAO-B)
Cimoxatone 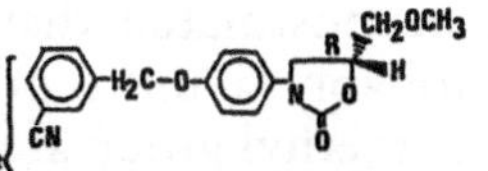	0.001	0.096	96	
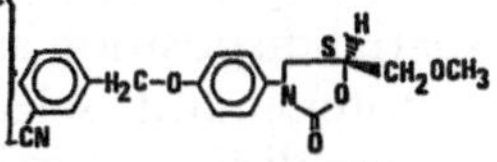	0.014	0.139	10	
Almoxatone 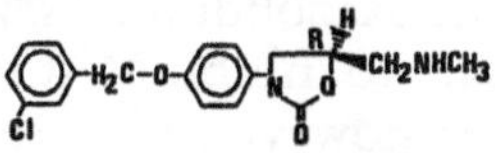	1.77	0.28		6.3
MD 240931 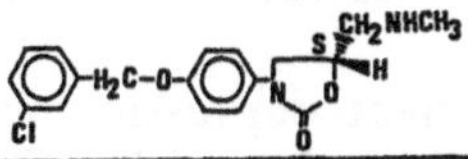	0.38	0.17		2.2

No preincubation of the inhibitor with the enzyme

Fig. 6. Ki values for MAO-A and -B of the enantiomers of cimoxatone, of almoxatone and of the corresponding S enantiomer

Replacement of the alcohol (Fig. 3, R = OH) by an ether or an amino group increases substantially the inhibition of the A or B form, respectively (Fig. 6).

Again the more potent and selective MAO-A inhibitor has the R configuration. In the case of MAO-B inhibitors, the more potent has the S configuration and the more selective the R configuration, which apparently seems to be the opposite to what was observed with the alcohol derivatives, but the R and S enantiomers in the alcohol and ether series have the same absolute configuration as the S and R enantiomers in the amine series. As discussed for alcohols, MAO-A is sensitive to the stereoisomerism of the β centre of ethers whereas MAO-B is not. This particular stereoselectivity of the A form is not confirmed with the amine derivatives.

Almoxatone was shown to be a substrate of both forms of MAO, whereas its S enantiomer seems to be metabolized essentially by the A form (Strolin Benedetti et al., 1983). It can therefore be postulated that both enantiomers are oxidized to the imine form but that a nucleophilic site present in the B form of the enzyme can react only with the imine form of the S enantiomer to give a covalent bond producing an irreversible inhibition.

The influence of the substitution of one of the hydrogens on the α carbon of cimoxatone by a methyl group on the inhibition of MAO-A has been investigated (Strolin Benedetti and Dostert, 1985). Cimoxatone and the erythro enantiomers of the α-methylcimoxatone [R (β), S (α); S (β), R α)] have the same MAO-A inhibitory potency, whereas the threo enantiomers were practically inactive. As it was shown that the R enantiomer of cimoxatone is a more potent MAO-A inhibitor than the S one, it can be postulated that the R (β), S (α) enantiomer of the erythro is probably a more potent inhibitor than the S (β), R (α) enantiomer. The methyl group at the α carbon in the R, S enantiomer has the same spatial disposition as the methyl group in (+)amphetamine, more potent and selective MAO-A inhibitor than the (−)enantiomer. In oxazolidinone series, a methyl group on the α carbon is not necessary to inhibit MAO-A, as it is in amphetamine comparatively to PEA; however, if present, it must be in S configuration.

The efficacy of MAO-A inhibitors to treat depression and of l-deprenyl to treat Parkinson's disease in combination with L-Dopa is presently recognized. In one study, treatment by L-Dopa combined with l-deprenyl prolonged life span significantly when compared with treatment by L-Dopa alone (Riederer et al., 1984). l-Deprenyl is a peculiar MAO-B inhibitor because of other pharmacological actions, mainly those resulting from its amphetamine-like sympathomimetic properties (Porsolt et al., 1984; Fozard et al., 1985). However, it seems that the amphetamine-type metabolites of l-deprenyl do not contribute to the therapeutic benefits conferred by this drug in parkinsonism (Elsworth et al., 1982). It follows that it would be of interest to study pure MAO-BIs in Parkinson's disease. Recently, Fozard et al. (1986) have demonstrated that selective inhibition of MAO-B by MDL 72145 increases the central effects of L-Dopa without modifying its cardiovascular effects. Their data provide no reason to suppose that MDL 72145 or other MAO-B inhibitors would be very different from l-deprenyl as an adjunct ot the L-Dopa-based therapy of Parkinson's disease.

Interesting new perspectives for l-deprenyl and other MAO-B inhibitors have recently appeared in the literature. Thus, l-deprenyl is under investigation in parkinsonian patients without L-Dopa but associated with bromocriptine (Calzetti and Baratti, 1985). Moreover, a recent publication refers to testing of a combined treatment with deprenyl and vitamin E, a general antioxidant, in Parkinson's disease (Lewin, 1985). The supposition that free radicals might be involved in cytotoxicity in parkinsonism is the rationale for suggesting antioxidant therapy, an example of which is vitamin E. In addition MAO might be involved in the generation of free radicals, hence a second rationale for testing the effect of inhibiting these enzymes to some degree. Free radicals produce lipid peroxidation. Very recently we have found a partial protection by l-deprenyl (10 mg/kg, p.o.), but not by clorgyline (10 mg/kg, p.o.), against in vivo lipid peroxidation, using ethane exhalation in CCl_4 treated rats as a model (Richard *et al.*, in press).

l-Deprenyl seems also to be active in some types of depression. Following a preliminary observation on the euphoriant and drive enhancing effect of l-phenylalanine in depressed patients with parkinsonism and encouraging results with d- or dl-phenylalanine in depressed patients, deprenyl has been co-administered with l-phenylalanine to unipolar depressed patients with a high percentage of beneficial effects (Birkmayer *et al.*, 1984).

References

Alston TA (1981) Suicide substrates for mitochondrial enzymes. Pharmac Ther 12: 1—41

Ask AL, Fagervall I, Florvall L, Ross SB, Ytterborn S (1986) Selective inhibition of monoamine oxidase by p-aminosubstituted phenylalkylamines in catecholaminergic neurones. Neuropharmacology 25: 33—40

Belleau B, Moran J (1963) Deuterium isotope effects in relation to the chemical mechanism of monoamine oxidase. Ann NY Acad Sci 107: 822—839

Birkmayer W, Riederer P, Linauer W, Knoll J (1984) L-deprenyl plus L-phenylalanine in the treatment of depression. J Neural Transm 59: 81—87

Biziere K, Bougault I, Kan JP, Mouget-Goniot C, Mons G, Worms P (1985) Effect of two amino-pyridazine derivatives on both forms of monoamine oxidase in rat brain. Br J Pharmacol 86 [Suppl]: 417 P

Boucher T, Strolin Benedetti M, Dostert P, Tipton KF (1987) IInd amine oxidase meeting, Uppsala, 2—4 August 1986. Acta Pharmacol Toxicol (in press)

Buckman TD, Sutphin MS, Eiduson S (1984) Proteases as probes of mitochondrial monoamine oxidase topography in situ. Mol Pharmacol 25: 165—170

Calzetti S, Baratti M (1985) A double-blind controlled study of the short-term effect of deprenyl (selegiline) combined with bromocriptine in early Parkinson's disease. 8th international symposium of Parkinson's disease, 9—12 June 1985, New York (abstract book, p 122)

Cesura AM, Kettler R, Da Prada M (1985) Binding of ^{3}H-Ro 16-6491 in human brain and platelets: characterization of MAO-B. J Neurochem 44 [Suppl]: S 179

Chiba K, Trevor A, Castagnoli N Jr (1984) Metabolism of the neurotoxic tertiary amine, MPTP, by brain monoamine oxidase. Biochem Biophys Res Commun 120: 574—578

Ciba-Geigy (1986) Sercloremine hydrochloride. Drugs of the Future 11: 126—128

Collins MA, Neafsey EJ (1985) β-carboline analogues of N-methyl-4-phenyl-1, 2, 5, 6-tetrahydropyridine (MPTP): endogenous factors underlying idiopathic Parkinsonism? Neurosci Lett 55: 179—184

Da Prada M, Kettler R, Keller HH, Bonetti EP, Imhof R (1986) Ro 16-6491: a new reversible and highly selective MAO-B inhibitor protects mice from the dopaminergic neurotoxicity of MPTP. In: Yahr MD, Bergmann KJ (eds) Adv neurol, vol 45. Raven Press, New York, pp 175—178

Da Prada M, Kettler R, Keller HH, Kyburz E, Burkhard WP (1987) IInd amine oxidase meeting, Uppsala, 2—4 August 1986. Acta Pharmacol Toxicol (in press)

Denney RM, Fritz RR, Patel NT, Widen SG, Abell CW (1983) Use of a monoclonal antibody for comparative studies of monoamine oxidase B in mitochondrial extracts of human brain and peripheral tissues. Mol Pharmacol 24: 60—68

Dostert P, Strolin Benedetti M (1986) Nouveaux inhibiteurs de la monoamine oxydase. Actual Chim Thér 13ème série, Lavoisier, Paris, pp 269—287

Elsworth JD, Sandler M, Lees AJ, Ward C, Stern GM (1982) The contribution of amphetamine metabolites of (—)-deprenyl to its antiparkinsonian properties. J Neural Transm 54: 105—110

Fowler CJ, Ross SB (1984) Selective inhibitors of monoamine oxidase A and B: biochemical, pharmacological, and clinical properties. Med Res Rev 4: 323—358

Fozard JR, Zreika M, Robin M, Palfreyman MG (1985) The functional consequences of inhibition of monoamine oxidase type B: comparison of the pharmacological properties of L-deprenyl and MDL 72145. Naunyn Schmiedeberg's Arch Pharmacol 331: 186—193

Fozard JR, Palfreyman MG, Robin M, Zreika M (1986) Selective inhibition of monoamine oxidase type B by MDL 72145 increases the central effects of L-Dopa without modifying its cardiovascular effects. Br J Pharmacol 87: 257—264

Fuller RW, Hemrick SK (1978) Steric influence on inhibition of monoamine oxidase forms by 2,3-dichloro-α-methylbenzylamine. Res Commun Chem Pathol Pharmacol 20: 199—202

Fuller RW, Hemrick-Luecke SK (1985) Inhibition of types A and B monoamine oxidase by 1-methyl-4-phenyl-1, 2, 3, 6-tetrahydropyridine. J Pharmacol Exp Ther 232: 696—701

Kalir A, Sabbagh A, Youdim MBH (1981) Selective acetylenic "suicide" and reversible inhibitors of monoamine oxidase types A and B. Br J Pharmac 73: 55—64

Kan JP, Strolin Benedetti M (1981) Characteristics of the inhibition of rat brain monoamine oxidase in vitro by MD 780515. J Neurochem 36: 1561—1571

Keller HH, Kettler R, Keller G, Da Prada M (1986) Short-acting novel MAO-inhibitors: in vitro evidence for reversibility of MAO inhibition. Naunyn Schmiedeberg's Arch Pharmacol 332 [Suppl]: R 74

Kenney WC, Nagy J, Salach JI, Singer TP (1979) Structure of the covalent phenylhydrazine adduct of monoamine oxidase. In: Singer TP, Von Korff RW, Murphy DL (eds) Monoamine oxidase: structure, function and altered functions. Academic Press, New York, pp 25–37

Knoll J, Ecsery Z, Magyar K, Satory E (1978) Novel (–)deprenyl-derived selective inhibitors of B-type monoamine oxidase. The relation of structure to their action. Biochem Pharmacol 27: 1739–1747

Lesage A, Strolin Benedetti M, Rumigny JF (1984) High affinity binding site for (+)amphetamine in rat hypothalamus: fact or artefact? Neurochem Int 6: 283–286

Lesage A, Strolin Benedetti M, Rumigny JF (1985) Evidence that (+)[^{3}H]amphetamine binds to acceptor sites wich are not MAO-A. Biochem Pharmacol 34: 3000–3002

Lewin R (1985) Clinical trial for Parkinson's disease? Science 230: 527–528

Magyar K, Ecseri Z, Bernath G, Satory E, Knoll J (1979) Structure-activity relationship of selective inhibitors of MAO-B. In: Magyar K (ed) Monoamine oxidases and their selective inhibition. Pergamon Press Advances in pharmacological research and practice, vol 4, pp 11–21

McDonald IA, Palfreyman MG, Zreika M, Bey P (1986) (Z)-2-(2, 4-dichlorophenoxy)methyl-3-fluoroallylamine (MDL 72638): a clorgyline analogue with surprising selectivity for monoamine oxidase type B. Biochem Pharmacol 35: 349–351

Meller E, Friedman E, Schweitzer JW, Friedhoff AJ (1977) Tetrahydro-β-carbolines: specific inhibitors of type A monoamine oxidase in rat brain. J Neurochem 28: 995–1000

Meyerson LR, McMurtrey KD, Davis VE (1976) Neuroamine-derived alkaloids: substrate-preferred inhibitors of rat brain monoamine oxidase *in vitro*. Biochem Pharmacol 25: 1013–1020

Nelson DL, Herbet A, Petillot Y, Pichat L, Glowinski J, Hamon M (1979) (^{3}H)Harmaline as a specific ligand of MAO-A-I. Properties of the active site of MAO-A from rat and bovine brains. J Neurochem 32: 1817–1827

Palfreyman MG, McDonald IA, Bey P, Danzin C, Zreika M, Lyles GA, Fozard JR (1986) The rational design of suicide substrates of amine oxidases. Biochem Soc Trans 14: 410–413

Paul SM, Hulihan-Giblin B, Skolnick P (1982) (+)-Amphetamine binding to rat hypothalamus: relation to anorectic potency of phenylethylamines. Science 218: 487–490

Pearce LB, Roth JA (1984) Monoamine oxidase: separation of the type A and B activities. Biochem Pharmacol 33: 1809–1811

Porsolt RD, Pawelec C, Roux S, Jalfre M (1984) Discrimination of the amphetamine cue. Effects of A, B and mixed type inhibitors of monoamine oxidase. Neuropharmacology 23: 569–573

Rando RR, Eigner A (1977) The pseudoirreversible inhibition of monoamine oxidase by allylamine. Mol Pharmacol 13: 1005–1013

Richard C, Guichard JP, Strolin Benedetti M, Dostert P (1987) Effect of MAO inhibitors on *in vivo* lipid peroxidation. IInd amine oxidase meeting, Uppsala, 2–4 August 1986. Acta Pharmacol Toxicol (in press)

Riederer P, Jellinger K, Seemann D (1984) Monoamine oxidase and parkinsonism. In: Tipton KF, Dostert P, Strolin Benedetti M (eds) Monoamine
oxidase and disease. Prospects for therapy with reversible inhibitors.
Academic Press, London, pp 403—415

Riley TN, Brier CG (1972) Absolute configuration of (+)- and (—)-trans-
2-phenylcyclopropylamine hydrochloride. J Med Chem 15: 1187—
1188

Russel SM, Davey J, Mayer RJ (1979) The vectorial orientation of human
monoamine oxidase in the mitochondrial outer membrane. Biochem J
181: 7—14

Silverman RB (1983) Mechanism of inactivation of monoamine oxidase by
trans-2-phenylcyclopropylamine and the structure of the enzyme-inactivator adduct. J Biol Chem 258: 14766—14769

Silverman RB, Hoffman SJ (1980) Mechanism of inactivation of mitochondrial monoamine oxidase by N-cyclopropyl-N-arylalkyl amines. J Am
Chem Soc 102: 884—886

Silverman RB, Zieske PA (1986) Identification of the amino acid bound to
the labile adduct formed during inactivation of monoamine oxidase by
l-phenylcyclopropylamine. Biochem Biophys Res Commun 135: 154—
159

Singer TP (1979) Active site-directed, irreversible inhibitors of monoamine
oxidase. In: Singer TP, Von Korff RW, Murphy DL (eds) Monoamine
oxidase: structure, function and altered functions. Academic Press, New
York, pp 7—24

Singer TP, Salach JI (1981) Interaction of suicide inhibitors with the active
site of monoamine oxidase. In: Youdim MBH, Paykel ES (eds) Monoamine oxidase inhibitors—the state of the art. J Wiley, New York,
pp 17—29

Strolin Benedetti M (1984) Some pharmacological and clinical aspects of
reversible MAO inhibitors. In: Paton W, Mitchell J, Turner P (eds) Pro-
ceedings vol 2, IUPHAR 9th int congress of pharmacology, London,
1984. Macmillan, Basingstoke, pp 219—230

Strolin Benedetti M, Dostert P (1985) Stereochemical aspects of MAO
interactions: reversible and selective inhibitors of monoamine oxidase.
TIPS 6: 246—251

Strolin Benedetti M, Dow J, Boucher T, Dostert P (1983) Metabolism of the
monoamine oxidase-B inhibitor, MD 780236 and its enantiomers by
the A and B forms of the enzyme in the rat. J Pharm Pharmacol 35:
837—840

Waldmeier PC, Felner AE, Tipton KF (1983 a) The monoamine oxidase
inhibiting properties of CGP 11305 A. Eur J Pharmacol 94: 73—83

Waldmeier PC, Feldtrauer JJ, Stoecklin K, Paul E (1983 b) Reversibility of
the interaction of CGP 11305 A with MAO A in vivo. Eur J Pharmacol
94: 101—108

Westlund KN, Denney RM, Kochersperger LM, Rose RM, Abell CW (1985)
Distinct monoamine oxidase A and B populations in primate brain.
Science 230: 181—183

Yu PH, Bailey BA, Durden DA, Boulton AA (1986) Stereospecific deuterium substitution at the α-carbon position of dopamine and its effect on oxidative deamination catalyzed by MAO-A and MAO-B from different tissues. Biochem Pharmacol 35: 1027–1036

Authors' address: Dr. Margherita Strolin Benedetti, Laboratoires Fournier, Centre de Recherches de Daix, 50 rue de Dijon, F-21121 Fontaine-les-Dijon, France.

J Neural Transm (1987) [Suppl] 23: 121–138

MAO inhibitors in mental disease: their current status

J. H. Dowson

Department of Psychiatry, University of Cambridge Clinical School, Addenbrooke's Hospital, Cambridge, United Kingdom

Summary

Available MAOIs seem to be mainly indicated for the heterogeneous group of patients with depressive syndromes. Although groups of patients with all the recognized major subtypes of depression (including "endogenous depression") probably respond in varying degrees, MAOIs appear to be particularly indicated for out-patients with "neurotic depression" complicated by panic disorder or hysteroid dysphoria, which involves repeated episodes of depressed mood in response to feeling rejected. MAOIs can also be effective in several anxiety syndromes, in particular panic disorder. Other reports have claimed success in a variety of other syndromes including bulimia, anorexia nervosa, obsessive-compulsive neurosis, atypical facial pain and some other types of chronic pain, childhood attention deficit disorder and delusions of infestation by parasites. The nature of any underlying personality disorder is an important response variable and the assessment of personality should be encouraged in further studies. The development of new drugs raises the prospect of a range of MAOIs targeted at specific patient populations.

Tranylcypromine also merits further investigation as clinical experience suggests that it can produce a dramatic response in some patients with phenelzine-resistant disorders. This may be due, at least in part, to its amphetamine-like effects.

The euphoria, overactivity and increased appetite that accompanied the use of iproniazid in the treatment of tuberculosis led to its introduction as an antidepressant in 1957 (Crane, 1957; Kline, 1958). Subsequently it was found that this drug was an inhibitor of monoamine oxidase, and various other compounds which shared this property were introduced into psychiatric practice. However, side-effects marred the clinical interest in MAOIs in the early 1960s; iproniazid and other hydrazine compounds were associated with hepatic toxicity, and there were reports of hypertensive episodes, some leading to serious and even fatal cerebrovascular complications. This bred an understandable attitude of clinical caution; however, once the mechanism of these reactions was understood, namely the release of stored catecholamines by dietary tyramine or sometimes by other drugs, these reactions could be avoided if dietary and other restrictions were observed. Another problem which has discouraged the widespread use of MAOIs has been that, despite a recognition that these compounds can produce some excellent therapeutic responses, MAOIs developed a reputation of being of limited clinical value, partly because it has been difficult to predict a good response and partly because inadequate doses were often prescribed.

In the U.K., iproniazid, isocarboxazid and phenelzine are the hydrazine MAOIs which, together with tranylcypromine, are currently available to the clinician, although iproniazid and isocarboxazid are seldom prescribed. It is generally assumed that their clinical efficacy is related to MAO inhibition, but other actions may also be important; for example, the amphetamine-like effects of tranylcypromine.

MAO activity is located in the outer membrane of the mitochondrion and there are two main forms of the enzyme. Recent data from monoclonal antibody studies suggest that MAO-A is present in human catecholamine-containing neurones, while 5-HT-containing cell bodies possess MAO-B, as do platelets. However, it should be noted that *in vitro* studies of substrate specificity may not parallel *in vivo* metabolism, and that there are species differences in substrate specificity. MAO activity in glial cells may be important, and, in the human brain, dopamine is said to be predominantly metabolised by MAO-B. (Glover and Sandler, 1986). Some neurones, at least in the rat brain, may contain both main forms of MAO, and a semicarbazide-sensitive form has also been reported (Callingham, 1986). It has been suggested that antidepressant effect is correlated with MAO-A inhibition (Murphy, Lipper, Pickar *et al.*, 1981).

While some MAOIs, including phenelzine and tranylcypromine, inhibit both major forms of MAO, others have a selective effect (usually partially selective), on MAO-A or MAO-B, which is sometimes dose-related (Tollefson, 1983). Chlorgyline selectively inhibits MAO-A, while pargyline and deprenyl preferentially inhibit MAO-B. Deprenyl, which inhibits dopamine breakdown, is claimed to be a useful adjunct to the treatment of Parkinson's disease with L-DOPA and to prolong survival in patients with this condition. In an attempt to develop more powerful and less toxic MAOIs, several selective MAO-A inhibitors have been introduced, for example, moclobemide, cimoxotone, CGP11305A, BWA616U, amiflamine, brofaremine and toloxatone which is registered for use in France. These are relatively shortacting and it is hoped that this and other properties may be associated with a reduced likelihood of various side-effects (Pare, 1985). However, their clinical evaluation is at an early stage.

In clinical practice, an MAOI seems to be mainly indicated for the heterogeneous group of patients with depressive syndromes, and most studies have employed phenelzine (Quitkin *et al.*, 1979). Research in this area is bedevilled with problems of syndrome classification, selection criteria and clinical assessment; also, many depressed patients spontaneously improve within several weeks, there are often non-specific environmental therapeutic factors which are difficult to control, and many early studies of phenelzine involved doses lower than about 1 mg/kg body weight per day which usually seems necessary for clinical response. Above a threshold dose level, the size of the dose for patients who respond may be related to the duration of the period before a significant clinical effect is produced.

The classification of depressive syndromes, in an attempt to obtain aetiologically homogeneous patient groups, can involve discrete categories, or ratings over a number of dimensions, although it has been claimed that the statistical analysis of dimensional data may often involve unwarranted assumptions. When considering this confusing area it can be helpful to bear in mind that there appear to be three main categories of aetiological factors. Firstly, "endogenous" or "biological" aetiology involving abnormal brain function, which often requires a genetically determined vulnerability. The resulting depression may appear "out of the blue", but even if the disorder is precipitated by life-events, the nature or severity of the symptoms are often clearly out of the context of the patient's personality and environment, corresponding to the lay person's concept of disease. Of course, the definition of "abnormal brain function" is a complex problem, akin to the definition of hypertension. However, although an arbitrary line between normality and abnormality may have to be drawn at one point in the spectrum of hypertension, some cases of hypertension at the most abnormal end of the spectrum are clearly due to qualitatively distinct abnormal somatic function and there is good evidence that the same can be said for some cases of depression. The second category of aetiological factor is past and present adverse environmental situations, which often interact with the third category, which is personality disorder. Personality is reflected by patterns of behaviour, and when these are judged to be seriously maladaptive, the term "personality disorder" is used. Certain severe personality disorders have been shown to be associated with genetic factors, (Baron et al., 1985; Cadoret et al., 1985; Kendler et al., 1981), and may sometimes involve a biological predisposition to mood disorder, (Gunderson and Elliott, 1985), while the social and environmental problems which are secondary to personality disorder can lead to a depressed mood which appears, at least in part, understandable in this context. Such a model of the causation of depressive syndromes must, of course, be oversimplistic, but accords with the experience of many clinicians. Depression may be derived from a mixture of two or three aetiological categories, although one may predominate.

Most categorical systems of classification of depression are based on symptoms and these can sometimes correlate with aetiology, as endogenous aetiology is associated with a varying number of specified clinical features which comprise the syndrome of so-called "endogenous depression", although these symptoms are not invariably present as markers of endogenous aetiology. The development of biological markers of endogenous aetiology, such as time from sleep onset to the first period of rapid eye movements, and the dexamethasone suppression test, have been used to improve patient selection, and further advances in this area are to be expected (Ansseau et al, 1985; Schatzberg et al., 1985).

The most enduring classification of depression involves just two categories, endogenous (or psychotic) and neurotic (or reactive) (Monroe et al., 1985; Young et al., 1986), and despite its limitations the clinical validity of this approach has stood the test of time. However, these terms are unsatisfactory. This classification is based on clinical description and although the features of the endogenous syndrome correctly infer that there is endogenous aetiology, endogenous aetiology does not invariably present with the specified symptoms. The term "psychotic", as an alternative to endogenous, is also misleading as it infers the presence of delusions or hallucinations; however these are not always part of an endogenous syndrome which rarely presents with all or even most of the various specified features. Finally, the term "reactive", as an alternative to neurotic, makes an aetiological claim which is not always justified; while the nature and severity of neurotic depression can often seem understandable (or meaningful) as a reaction of that individual to events, endogenous aetiology may also contribute to the development of a neurotic depression which would be reflected by a syndrome which was incompletely understandable. Other classifications of depression have been based on the presence or absence of a history of mania, of another non-affective psychiatric disorder, or of delusions (Charney and Wilson, 1981).

Many studies have examined the clinical features associated with the endogenous syndrome; these include marked agitation or psychomotor retardation with slowed thoughts and actions; absence of mood reactivity to pleasant circumstances with a loss of interest in pleasurable activity; self-reproach; morning worsening; early morning wakening; marked loss of appetite and weight; a distinct quality of mood which is unlike normal unhappiness; hallucinations and delusions (Nelson and Charney, 1981). However, the number and severity of these features vary considerably. In patients with features of the endogenous syndrome there is less likely to be an identifiable precipitating stress compared with those with neurotic depression, but there are exceptions, as it seems that life events can sometimes precipitate the biological processes underlying the endogenous syndrome.

Neurotic depression encompasses a heterogeneous group of clinical presentations which tend to be less severe than endogenous depressions, but again there are exceptions. There is a varying and inadequately defined number of a long list of symptoms including anxiety, somatic concomitants of anxiety and phobias, in addition to depressed mood. Most patients with a depressive syndrome fall within the category of neurotic depression, but progress has been made in identifying clinically valid sub-types which reflect differences in aetiology. For example, Akiskal (1983), identified four subtypes of neurotic depression; the first, primary depression with residual chronicity, occurs in an individual with previously normal mood and appears to be a syndrome with endogenous aetiology which usually responds to tricyclic antidepressants. The second sub-type, subaffective dysthymic disorder, is claimed to be an aetiologically similar syndrome which, as it occurs before the age of 25, can appear to be a personality disorder, as the patient may have appeared chronically depressed throughout adult life, often presenting as quiet, passive, gloomy, self-critical and conscientious. This disorder is also said to respond to tricylic antidepressants. The third sub-type is chronic secondary dysphoria and is a chronic neurotic depressive syndrome which seems understandable as a reaction to some other chronic medical condition such as a severe deformity. Akiskal suggests that some of these patients respond to an MAOI. Finally, there are the character-sprectrum disorders in which the patient is chronically depressed since childhood or adolescence in the context of various types of severe personality disorder, for example with dependent or sociopathic traits, which are often associated with self-injury and antisocial behaviour. Akiskal claims that such patients do not respond significantly to antidepressant medication, a view which would be endorsed by many clinicians.

To return to the classification of depressive syndromes into endogenous and neurotic, there is evidence that some patients with either syndrome can respond to an MAOI, but although MAOIs may be preferentially effective in certain patients, a tricyclic anti-depressant has tended to be the treatment of first choice for a depressive syndrome if medication is indicated. However, a more flexible attitude is emerging.

Many reports have examined clinical predictors of response to MAOIs within the neurotic depression group. (Liebowitz et al., 1984; Nies and Robinson, 1982; Pare, 1985; Tollefson, 1983), and the term "atypical depression" has been used to indicate the features which have been considered particularly suitable for MAOI treatment (Davidson et al., 1982). This term has been used in two ways; firstly to denote neurotic depression complicated by marked anxiety, including phobias, and, secondly, in relation to the presence of certain physiological changes which are opposite to those in the endogenous depressive syndrome; the patient feels worse in the evening, has difficulty in getting off to sleep or sleeps excessively, and has an increased appetite.

To summarize an extensive literature, the following features have been claimed to predict a favourable response of a neurotic depressive syndrome to MAOIs: maintenance of mood reactivity, various forms of anxiety including panic attacks, overeating, fatigue, evening worsening and "rejection sensitivity". The latter is a central feature of so-called "hysteroid dysphoria" which may respond preferentially to MAOIs and involves repeated episodes of depressed mood in response to feeling rejected, in the setting of certain personality traits (Liebowitz and Klein, 1979).

Patients with neurotic depression may often present with complaints of somatic anxiety and this can involve so-called "cardiac neurosis" which is a hypochondriacal preoccupation with cardio-vascular function, and it has been claimed that this can often benefit from an MAOI. A good response is also said to be associated with a lack of guilt and self-reproach and the absence of severe forms of personality disorder.

A study by Rowan and colleagues (1982) examined the effects of adequate doses of phenelzine, amitriptyline and placebo in a large group of out-patients with neurotic depression. Both drugs were found to be of similar efficacy and clearly superior to placebo. Improvement began at about two weeks but maximum effect was not gained until after about six weeks. Although this study confirmed the value of phenelzine in the traditional target group of out-patients with neurotic depression, it raised the question whether the supposed difference between the antidepressant effects of MAOIs and tricyclics in this population has been exaggerated. However, as will be described, there is evidence that certain sub-types of neurotic depression may show a preferential response ot phenelzine compared with a tricyclic, and it has been suggested that in certain depressed patients a lack of response to a tricyclic antidepressant can be a predictor of a response to an MAOI.

The evidence of the efficacy of MAOIs in the endogenous syndrome is conflicting (Davidson *et al.*, 1981; Murphy *et al.*, 1981; Tollefson, 1983; Pare, 1985; Paykel *et al.*, 1979), but positive results have been claimed, particularly in association with psychomotor retardation and the absence of delusions. The general view seems to be that an MAOI can be effective, and perhaps preferentially effective, in individual patients with an endogenous syndrome, but, in groups, tricyclic antidepressants are better. Therefore, an MAOI is seldom a first choice for the treatment of an endogenous depressive syndrome.

Many studies report good responses to phenelzine or tranylcypromine in so-called "treatment-resistant" depression, which may involve either an endogenous syndrome or a neurotic syndrome which has failed to respond to a tricyclic antidepressant or electroconvulsive therapy (ECT) (Pare, 1985). Many of these are in the elderly and there is an increasing awareness that MAOIs can benefit many elderly depressed patients (Georgotas *et al.*, 1981). Also, treatment-resistant endogenous syndromes can appear to respond to a combination of an MAOI with other drugs, such as lithium, L-tryptophan or a tricyclic antidepressant (Feighner *et al.*, 1985; Price *et al.*, 1985), and there are several uncontrolled case reports which suggest that chlorgyline may benefit some patients with a rapid cycling bipolar manic-depressive disorder which does not respond to standard treatment with lithium (Potter *et al.*, 1982).

In summary, MAOIs seem effective in a range of depressive syndromes, but there is most evidence of efficacy in relatively mild neurotic depressive syndromes found in out-patient populations. Tricyclic antidepressants are also effective in this heterogeneous group and some studies have failed to show a preferential effect of an MAOI. However, there is also evidence that the following sub-types of neurotic depression may respond preferentially to MAOIs: those with marked anxiety, including panic attacks and phobias, those with

Akiskal's chronic secondary dysphoria involving depression in the context of another disabling illness, and those with hysteroid dysphoria which occurs in patients with hysterical personality traits involving child-like over-dependence, egocentricity and emotional lability. Personality and personality disorder in relation to MAOI response will now be considered in more detail.

Klein (1977) has described the features which comprise the recently delineated syndrome of hysteroid dysphoria. Women are usually affected and there are repeated episodes of depressed mood as a consequence of feeling rejected usually within close relationships. Such individuals are over-reliant on approval, attention or praise, and show an extreme intolerance to personal rejection. They have been called "attention junkies". Mood reactivity is preserved and fatigue is a common complaint together with overeating and over-sleeping. Liebowitz and colleagues (1984) reported a trial of phenelzine or impramine or placebo in patients with neurotic depression, and analysed the results within three subgroups; patients with depression and panic attacks, those with depression and hysteroid dysphoria and those with depression alone. The results suggested that a history of panic attacks or hysteroid dyshoria were predictive of a preferential response to phenelzine, although the evidence in relation to hysteroid dysphoria was tentative. However, a study by Kayser and colleagues (1985) provides further evidence of Klein's original claim that hysteroid dysphoria preferentially responds to an MAOI when compared with a tricyclic antidepressant. A questionnaire eliciting features of hysteroid dysphoria was administered to depressed out-patients and 47 completed a 6-week double-blind study comparing the effects of phenelzine with amitriptyline. A high scoring hysteroid dysphoria sub-group was identified, and 9 out of the 9 high scorers treated with phenelzine responded, in contrast to only 3 out of 5 high scorers treated with amitriptyline. Of course the numbers treated with amitriptyline were small and further studies are required. Although the efficacy of phenelzine in hysteroid dysphoria seems clear the value of tricyclics is less certain.

Tyrer and colleagues (1983) carried out an important study of the association between underlying personality disorder, neurotic depression and phenelzine response. 316 consecutively-referred out-patients had a variety of neurotic syndromes and 144 had a diagnosis of neurotic depression. Personality disorders of various types were judged to be present in 40% of the sample and, in general, a pre-existing personality disorder was associated with a poor phenelzine response; obsessional personality involving excessive rigidity, orderliness and indecision, was the most common personality disorder and this personality syndrome merits attention in further studies.

In summary, severe personality disorder of several types, for example obsessional, hysterical, schizoid, borderline and socio-pathic, seems to be a predictor of a poor response of neurotic depres-sion to MAOIs. Schizoid personality involves emotional unrespon-siveness and social isolation while borderline personality denotes impulsive emotional outbursts with selfdamaging or antisocial behaviour, often involving drug or alcohol abuse. Such individuals are angry, demanding, and miserable and have difficulty in living unsupported between a succession of stormy personal relationships (Gunderson and Elliott, 1985). However, neurotic depression in the context of a relatively mild hysterical personality syndrome, which can present with hysteroid dysphoria, may often respond well to an MAOI. The relevance of other less severe personality syndromes to MAOI response requires further investigation.

Anxiety is a common ingredient in neurotic depression and is a response-predictor of this syndrome to MAOIs, but MAOIs are also effective when anxiety, in particular panic attacks, is the primary dia-gnosis (Nies and Robinson, 1981; Sheehan et al., 1980; Sheehan, 1984). While anxiety can often seem psychologically understandable as an interaction of the individual with the environment, evidence is accumulating of endogenous aetiology in some anxiety syndromes, particularly in relation to the spontaneous episodes of overwhelming fear known as panic attacks. These can occur alone or in combina-tion with other disorders, in particular agoraphobia. The evidence suggests that both MAOIs and tricyclics can be of benefit for anxiety disorders, but panic attacks may respond preferentially to MAOIs. Further comparative studies are required.

Although MAOIs have been mainly targeted at depression or anxiety, there are many other reports of their use in a variety of less common psychiatric disorders, such as bulimia, anorexia nervosa, obsessive-compulsive neurosis, atypical facial pain and some other forms of chronic pain, childhood attention deficit disorder, depres-sion in the context of neurological disease and certain isolated delusions such as having an abnormal facial feature or being infested with parasites (Hudson et al., 1985; Jenike, 1984, 1985; Krishnan et al., 1984).

Bulimia consists of binge eating with self-induced vomiting, and usually occurs in women whose weight is within normal limits. In a double-blind placebo-controlled trial of phenelzine in 20 patients, 5 out of the 9 in the treatment group ceased bingeing while none stopped in the placebo group (Walsh et al., 1984).

Most of the reports of MAOIs in obsessive-compulsive disorder, which involves recurrent or persistent thoughts, and behavior such as excessive cleaning, checking and avoidance, are uncontrolled, and a study comparing chlorgyline with the standard drug treatment clomipramine did not indicate any therapeutic effect of the MAOI (Zann et al., 1984). However several case reports have claimed a dramatic response to tranylcypromine (Jenike, 1981; Swinson, 1984).

The role of certain stimulants in the management of some hyperactive children has been established, and in a double-blind cross-over study of chlorgyline, tranylcypromine and dextroamphetamine, it appeared that the MAOIs had equivalent efficacy to dextroamphetamine, which is one of the standard treatments, resulting in an improvement in restless, disruptive behaviour (Zametkin et al., 1985).

Various studies have investigated the use of several MAOIs which are not generally available. The selective MAO-A inhibitor chlorgyline has been claimed to have a greater antidepressant effect than the partially selective MAO-B inhibitor pargyline (Lipper et al. 1979), but chlorgyline probably has no fewer sideeffects than phenelzine (Pare, 1985). It was hoped that a selective MAO B inhibitor such as deprenyl would be less toxic and, indeed, this drug does not seem to cause orthostatic hypotension or (if the dose is not large enough to lose B selectivity), tyramine-induced hypertension, but despite various claims its antidepressant effect is uncertain (Elsworth et al., 1978; Karoum et al., 1982; Mendlewicz and Youdim, 1980; Quitkin et al., 1984). Mann and colleagues (1984) have reported that deprenyl may have an antidepressant effect in neurotic syndromes in the absence of panic attacks and phobias, but of 3 placebo-controlled studies, one was negative, while another noted that better results were produced by relatively high doses which may also have inhibited MAO-A (Aarons et al., 1984; Mendis et al., 1981 a, b; Mendlewicz and Youdim, 1983).

The evaluation of the newer selective MAO-A inhibitors such as moclobemide is also at an early stage. An Italian study (Casacchia et al., 1984) using a double-blind placebo design, claimed a marked antidepressant effect within four weeks, but the characteristics of the patient group are unclear. Another study by Norman and colleagues (1985) compared the effects of amitriptyline and moclobemide in a heterogeneous group of depressed in-patients and found no difference between the improvement associated with both drugs. A similar study by Larsen and colleagues (1984) compared moclobemide with clomipramine in patients whose characteristics are unclear but who probably had neurotic depressive syndromes. Both studies have claimed that moclobemide and the tricyclic antidepressant were equally effective, but in the absence of a control group this conclusion cannot be justified in view of the possibility of spontaneous improvements and probable placebo effects. In another uncontrolled study (Dajas *et al.*, 1984) it was claimed that moclobemide can benefit so-called major depressive disorders, many of which would probably attract the diagnosis of endogenous depression.

Although the clinical efficacy of moclobemide is still uncertain, there is evidence that it is less prone to tyramine-induced hypertension when compared with standard MAOIs (Korn *et al.*, 1984) and after use in over 1000 patients there is no evidence of hepatotoxicity. Also, moclobemide is claimed to have fewer side-effects when compared with tricyclic antidepressants and the standard dose can usually be given from the outset, which may help to minimize the delay before the appearance of any therapeutic effect. It has been claimed that the risk of a reaction to an inadvertent tyramine-rich meal, involving as much as 100 mg of tyramine, is minimal if moclobemide is routinely taken after meals, but the search goes on for an effective MAOI devoid of any risk of sympathomimetic crisis. It is possible that a combination of a selective MAO-A inhibitor with another drug inhibitor, or the development of brain-selective MAOIs might completely eliminate this problem.

Future studies on new and old MAOIs should pay particular attention to the likely aetiological homogeneity of patient groups, and as aetiology is not always related to symptoms, selection criteria should, if possible, include biological markers and personality variables. Also it is important to use a standardized and comprehensive method of clinical assessment such as Paykel's Clinical Interview (Paykel, 1985). Unlike many commonly-used instruments this is suitable for the whole range of depressive syndromes. There is a need to study the effects of MAOIs in non-clinical populations, and the use of MAOIs and drug combinations in responder groups.

Although most studies have involved phenelzine, there is little evidence as to the relative efficacy of the various MAOIs, although clinical experience suggests that the choice of drug may be important. Antidepressants should not be prescribed routinely for every patient with a relatively mild depressive syndrome, as many will remit spontaneously or respond to a psychotherapeutic approach alone. An adequate dose must be prescribed and there is evidence that this should inhibit at least 85% of platelet MAO activity (Murphy *et al.*, 1984). Usually 60—90 mg/day of phenelzine is required. Mild orthostatic hypotension, as well as decreased lying systolic blood pressure usually occurs after about 14 days and the maximum effect on blood pressure is seen after about 4 weeks (Kronig *et al.*,

1983). This can be used as an estimate of the adequacy of the dose in the absence of laboratory facilities for platelet MAO estimation (Pare, 1985).

Many publications have considered the rate of acetylation of phenelzine, as this appears to be a major route of the drug's metabolism, despite a claim to the contrary (Cooper et al., 1984). The acetylator status is genetically determined and so-called slow and fast acetylators have been identified although differences vary in different countries. Three studies of phenelzine in depressed patients have claimed that slow acetylators showed more improvement compared with fast acetylators, although this was not found in five other investigations (Pare, 1985). At present, the clinical implications of acetylator status are uncertain; any differences in response may be short-lived and unimportant at the dose levels which are recommended.

The possible effects of intermittent rather than continuous treatment may be important (Pickar et al., 1982), and it has been suggested that an intermittent regime may be useful for some treatment-resistant depressions and to reverse the development of tolerance. If this is the case, the optimal time interval between drug courses, and the length of the courses, will require investigation.

Common side-effects include insomnia, with both increased wakening and decreased total sleep time, daytime drowsiness, increased appetite, weight gain and impaired sexual performance. Paradoxically, sleep may occasionally improve if the dose is given in the latter part of the day, and the addition of L-tryptophan or a benzodiazepine at night may also help (Nies and Robinson, 1982). An MAOI should be discontinued gradually to minimize various withdrawal symptoms including nocturnal myoclonic movements and nightmares, and the problem of dependence on MAOIs merits further study. If the dietary and other drug restrictions are observed, the risk of hypertension, headache, and cardiac dysrhythmias is generally considered to be negligible, although clinicians must guard against complacency and non-compliance. It has been claimed that only mature cheese, pickled fish, concentrated yeast extract and broad bean pods warrant absolute prohibition (Sullivan and Shulman, 1984), however, wider dietary restrictions are usually recommended (McCabe and Tsuang, 1982). It is of interest that the combination of a tricyclic antidepressant with an MAOI appears to afford significant protection against the tyramine effect, but only in some patients (Pare et al., 1984).

MAOIs, and also tricyclic antidepressants, can induce various behavioural disorders, in particular mania, and delirium can occur, especially in the elderly. Pickar and colleagues (1982) looked at the development of mania, with pressured speech and overactivity, in relation to phenelzine, chlorgyline and pargyline in 39 in-patients. Five developed mania and there was no clear difference between the drugs, although the numbers were small. It is unclear whether a genetically-determined vulnerability to manic-depressive disorder is usually necessary for this complication to develop, but an association with a previous history of mania was found. There was a delay in onset for at least 18 days, with the mean at about 3 weeks, and a similar latent period has been reported for the induction of mania by tricyclic antidepressants. However mania can occur after only a few days. It has been previously reported that the risk of hypomania or mania is about 11% with MAOIs (tranylcypromine may have a greater risk than phenelzine) and about 8% for tricyclics. It was found that a history of previous courses of MAOIs was associated with a greater severity of drug-induced mania and a lesser period before its onset, while in three patients mania developed after the drug was discontinued. In long-term use, MAOIs and indeed other psychotropic drugs may induce important modifications in receptor sensitivity accompanied by complex adaptational changes; these may be related to certain therapeutic effects, to certain side-effects and to tolerance. A rare but serious complication of a number of drugs is leukopenia; although only one case has been published in association with phenelzine, five other cases have been notified (Tipermas et al., 1984).

Clinical experience suggests that the combination of an MAOI with a tricyclic antidepressant can benefit some patients with "treatment-resistant" depressions although there is no evidence of this from controlled trials (Ananth and Luchins, 1977; Pare, 1985). After a washout period both drugs may be started together at low doses which are gradually increased. Certain combinations are dangerous, for example clomipramine should be avoided. Toxicity can involve hyperthermia, delirium and convulsions, so careful supervision is essential (White and Simpson, 1984). Other drugs which have been combined with an MAOI include L-tryptophan (this is probably the most commonly-used combination), stimulants, triiodothyronine and lithium; it is likely that these can sometimes be effective, but more evidence is required (Baldessarini, 1984; Feighner et al., 1985; Hullett and Bidder, 1983; Price et al., 1985; Tollefson, 1983).

It has been claimed that prophylactic treatment with MAOIs may be required for months or years, and that this can sometimes be achieved with a reduction of the initial target dose, even to as low as 15 mg of phenelzine per day. However, some patients relapse if the MAOI is reduced or discontinued. Finally, in the consideration of the use of any drug, the problem of compliance should not be forgotten (Tollefson, 1983).

In conclusion, despite the fact that MAOIs have not yet achieved an established place in psychiatric practice, it appears that available MAOIs can be effective, particularly for outpatients with depressive syndromes complicated by panic disorder or hysteroid dysphoria, in whom the effects may be greater than that of a tricyclic antidepressant. Studies in groups of patients may disguise the preferential and dramatic effects which seem to occur in many individual patients. The nature of any underlying personality disorder is often a significant response variable and the assessment of personality should form an important part of further studies. The development of new drugs raises the prospect of a range of MAOIs with fewer side-effects targeted at specific patient populations.

Tranylcypromine also merits further study, as clinical experience suggests that it can produce a dramatic response in some patients with phenelzine-resistant disorders, which may be due, in part, to the significant amount of amphetamine which is produced in its breakdown. Perhaps the use of tranylcypromine is re-evaluating the use of the pre-1957 stimulant antidepressants in psychiatric practice, and Cookson and Silverstone (1986) have recently confirmed that methylamphetamine can produce a marked elevation of depressed mood.

References

Aarons SF, Mann JJ, Frances A, Kocsis J, Young RC (1984) Antidepressant effect of L-deprenyl: relative contribution of MAO-A versus MAO-B inhibition. Abstract, collegium international, neuro-psychopharmacologium, 14th C.I.N.P. congress, Florence, June 1984

Akiskal HS (1983) Dysthymic disorder: psychopathology of proposed chronic depressive subtypes. Am J Psychiat 140: 11–20

Ananth J, Luchins D (1977) A review of combined tricyclic and MAOI therapy. Comp Psychiat 18: 221–230

Ansseau M, Kupfer DJ, Reynolds CF (1985) Internight variability of REM latency in major depression: implications for the use of REM latency as a biological correlate. Biol Psychiat 20: 489–505

Baldessarini RJ (1984) Treatment of depression by altering monoamine metabolism: precursors and metabolic inhibitors. Psychopharmacol Bull 20: 224–239

Baron M, Gruen R, Asnis L, Lord S (1985) Familial transmission of schizotypal and borderline personality disorders. Am J Psychiat 142: 927–934

Cadoret RJ, O'Gorman TW, Troughton E, Heywood E (1985) Alcoholism and antisocial personality. Interrelationships, genetic and environmental factors. Arch Gen Psychiat 42: 161–167

Callingham BA (1986) Some aspects of monoamine oxidase pharmacology. Cell Biochem Funct 4: 99–108

Casacchia M, Carolei A, Barba C (1984) A placebo-controlled study of the antidepressant activity of moclobemide, a new MAO-A inhibitor. Pharmacopsychiatry 17: 122–125

Charney DS, Nelson JC (1981) Delusional and nondelusional unipolar depression: further evidence for distinct subtypes. Am J Psychiat 138: 328–332

Cookson J, Silverstone T (1986) The effects of methylamphetamine in mood and appetite in depressed patients: A placebo-controlled study. Int Clin Psychopharmacol 1: 127–133

Cooper TB, Juidal SP, Robinson DS, Corcella J (1984) Metabolism of phenelzine in man: Lack of evidence for acetylation pathway. Abstract, collegium international neuro-psychopharmacologium, 14th C.I.N.P. congress, Florence

Crane GE (1957) Iproniazid (Marzilid) phosphate, a therapeutic agent for mental disorders and debilitating disease. Psychiat Res Rep 8: 142–152

Dajas F, Lista A, Barbeito L (1984) High urinary norepinephrine excretion in major depressive orders: Effects of a new type of MAO inhibitor moclobemide, Ro 11–1163). Acta Psychiat Scand 70: 432–437

Davidson J, McLeod MMN, Turnbull CD, Miller RD (1981) A comparison of phenelzine and imipramine in depressed inpatients. J Clin Psychiat 42: 10

Davidson JRT, Miller RD, Turnbull CD, Sullivan JL (1982) Atypical Depression. Arch Gen Psychiat 39: 527–534

Elsworth JD, Glover V, Reynolds GP (1978) Deprenyl administration in man: A selective monoamine oxidase B inhibitor without the "cheese effect". Psychopharmacology 57: 33–38

Feighner JP, Herbstein J, Damlouji N (1985) Combined MAOI, TCA and direct stimulant therapy of treatment-resistant depression. J Clin Psychiat 46: 206–209

Georgotas A, Mann J, Friedman E (1981) Platelet monoamine oxidase inhibition as a potential indicator of favourable response to MAOIs in geriatric depressions. Biol Psychiat 16: 997–1001

Glover V, Sandler M (1986) Clinical chemistry of monoamine oxidase. Cell Biochem Function 4: 89–97

Gunderson JG, Elliott GR (1985) The interface between borderline personality disorder and affective disorder. Am J Psychiat 142: 277–288

Hudson JI, Pope HG, Jonas JM, Yurgelun-Todd D (1985) Treatment of anorexia nervosa with antidepressants. J Clin Psychopharmacol 5: 17–23

Hullett FJ, Bidder TG (1983) Phenelzine plus triiodothyronine combination in a case of refractory depression. J Nerv Ment Dis 171: 318–320

Jenike MA (1981) Rapid response of severe obsessive-compulsive disorder to tranylcypromine. Am J Psychiat 138/9: 1249–1250

Jenike MA (1984) A case report of successful treatment of dysmorphophobia with tranylcypromine. Am J Psychiat 141: 1463–1464

Jenike MA (1985) Monoamine oxidase inhibitors as treatment for depressed patients with primary degenerative dementia. Am J Psychiat 142: 763–764

Karoum F, Chuang LA, Eisler T (1982) Metabolism of (–) deprenyl to amphetamine and methamphetamine may be responsible for deprenyl's therapeutic benefit: A biochemical assessment. Neurology (NY) 32: 503–509

Kayser A, Robinson DS, Nies A, Howard D (1985) Response to phenelzine among depressed patients with features of hysteroid dysphoria. Am J Psychiat 142: 486–488

Kendler KS, Gruenberg AM, Strauss JS (1981) An independent analysis of the Copenhagen sample of the Danish adoption study of schizophrenia, II. The relationship between schizotypal personality disorder and schizophrenia. Arch Gen Psychiat 38: 982–984

Klein DF (1977) Pharmacological treatment and delineation of borderline disorders. In: Hartocollis P (ed) Borderline personality disorders: the concept, the syndrome, the patient. International Universities Press, New York, pp 365–383

Kline NS (1985) Clinical experience with iponiazid (Marsilid). J Clin Exp Psychopathol 19 [Suppl] 1: 72–78

Korn A, Eichler HG, Gasic S, Raffesberg W, Medizinische I (1984) Moclobemide, a new MAO-A inhibitor, does not provoke a "cheese reaction" after ingestion of tyramine containing meals. J Pharm Pharmacol 36 [Workshop Suppl]: 64

Krishnan RR, Davidson J, Miller R (1984): MAO inhibitor therapy in trichotillomania associated with depression: Case report. J Clin Psychiat 45: 267–268

Kronig MH, Roose SP, Walsh BT, Woodring S, Glassman AH (1983) Blood pressure. Effects of phenelzine. J Clin Pharmacol 3: 307–310

Larsen JK, Holm P, Mikkelson PL (1984) Moclobemide and desipramine in the treatment of depression. Acta Psychiat Scand 70: 254–260

Liebowitz MR, Klein DF (1979) Hysteroid dysphoria. Psychiat Clin Am 2: 555–575

Liebowitz MR, Quitkin FM, Stewart JW, McGrath PJ, Harrison W, Rabkin J, Tricamo E, Markowitz JS, Klein DF (1984) Phenelzine v imipramine in atypical depression. Arch Gen Psychiat 41: 669–677

Lipper S, Murphy DL, Slater S, Buchsbaum MS (1979) Comparative behavioral effects of chlorgyline and pargyline in man: a preliminary evaluation. Psychopharmacology 62: 123–128

Mann JJ, Aarons SF, Frances AJ, Brown RD (1984) Studies of selective and reversible monoamine oxidase inhibitors. J Clin Psychiat 45: 62–66

McCabe B, Tsuang MT (1982) Dietry consideration in MAO inhibitor regimens. J Clin Psychiat 43: 178–181

Mendis N, Pare CMB, Sandler M, Glover V, Stern GM (1981 a) Is the failure of (–) deprenyl, a selective monoamine oxidase B inhibitor, to alleviate depression related to freedom from the cheese effect? Psychopharmacology 73: 87–90

Mendis N, Pare CMB, Sandler M, Glover V (1981 b) (—)-Deprenyl in the treatment of depression. In: Youdim MBH, Paykel ES (eds) Monoamine oxidase inhibitors—the state of the art. J Wiley, London

Mendlewicz J, Youdim MBH (1980) Antidepressant potentiation of 5-hydroxytryptophan by L-deprenyl in affective illness. J Affect Dis 2: 137—146

Mendlewicz J, Youdim MBH (1983) L-Deprenyl, a selective monoamine, oxidase type B inhibitor, in the treatment of depression: a double blind evaluation. Br J Psychiat 142: 508—511

Monroe SM, Thase ME, Hersen M, Himmelhock JM, Bellak AS (1985) Life events and the endogenous-nonendogenous distinction in the treatment and posttreatment course of depression. Comp Psychiat 26: 175

Murphy DL, Lipper S, Pickar D, Jimerson D, Cohen RM, Garrick NA, Alterman IS, Campbell IC (1981) Selective inhibition of monoamine oxidase type A: clinical antidepressant effects and metabolic changes in man. In: Youdim MBH, Paykel ES (eds) Monoamine oxidase inhibitors—the state of the art. J Wiley, Chichester

Murphy DL, Sunderland T, Cohen RM (1984) Monoamine oxidase—inhibiting antidepressants. Psychiat Clin N Am 7: 549—562

Nelson JC, Charney DS (1981) The symptoms of major depressive illness. Am J Psychiat 138: 1—12

Nies A, Robinson MD (1981): Comparison of clinical effects of amitriptyline and phenelzine treatment. In: Youdim MBH, Paykel ES (eds) Monoamine oxidase inhibitors—the state of the art. J Wiley, Chichester

Nies A, Robinson DS (1982) Monoamine oxidase inhibitors. In: Paykel ES (ed) Handbook of affective disorders. Churchill Livingstone, Edinburgh

Norman TR, Ames D, Burrows GD, Davies B (1985) A controlled study of a specific MAO-A reversible inhibitor (Ro 11-1163) and amtriptyline in depressive illness. J Aff Dis 8: 29—35

Pare CMB (1985) The present status of monoamine oxidase inhibitors. Br J Psychiat 146: 576—584

Pare CMB, Mousarwi M Al, Sandler M, Glover V (1984) Oral tyramine pressor test in patients receiving a combination of an MAOI and a tricyclic antidepressant. J Pharm Pharmacol 36 [Workshop Suppl]: 65

Paykel ES (1985) The clinical interview for depression. Development, reliability and validity. J Aff Dis 9: 85—96

Paykel ES, Parker RR, Penrose RJJ, Rassaby ER (1979) Depressive classification and prediction of response to phenelzine. Br J Psychiat 134: 572—581

Pickar D, Murphy DL, Cohen RM, Campbell IL, Lipper S (1982) Selective and nonselective monoamine oxidase inhibitors. Arch Gen Psychiat 39: 535—540

Potter WZ, Murphy DL, Wehr TA, Linnoila M, Goodwin FK (1982) Chlorgyline. A new treatment for patients with refractory rapid-cycling disorder. Arch Gen Psychiat 39: 505—510

Price LH, Charney DS, Heninger GR (1985) Efficacy of lithium-tranylcypromine treatment in refractory depression. Am J Psychiat 142: 619—623

Quitkin FM, Liebowitz MR, Stewart JW (1984) L-deprenyl in atypcial depressives. Arch Gen Psychiat 41: 777–781

Quitkin F, Rifkin A, Klein DF (1979) Monoamine oxidase inhibitors. A review of antidepressant effectiveness. Arch Gen Psychiat 35: 749–760

Rowan PR, Paykel ES, Parker RR (1982) Phenelzine and amitriptyline: effects on symptoms of neurotic depression. Br J Psychiat 140: 475–483

Schatzberg AF, Rothschild AJ, Gerson B, Lerbinger JE, Schildkraut JJ (1985) Towards a biochemical classification of depressive disorders IX. Br J Psychiat 146: 633–637

Sheehan DV (1984) Delineation of anxiety and phobic disorders responsive to monoamine oxidase inhibitors: Implications for classification. J Clin Psychiat 45: 29–36

Sheehan DV, Ballenger J, Jacobsen G (1980) Treatment of endogenous anxiety with phobic, hysterical and hypochondriacal symptoms. Arch Gen Psychiat 37: 51–59

Sullivan EA, Shulman KI (1984) Diet and monoamine oxidase inhibitors: A re-examination. Can J Psychiat 29: 707–711

Swinson RP (1984) Response to tranylcypromine and thought stopping in obsessional disorder. Br J Psychiat 144: 425–427

Tipermas A, Gilman HE, Russakoff LM (1984) A case report of leukopenia associated with phenelzine. Am J Psychiat 141: 806–807

Tollefson GD (1983) Monoamine oxidase inhibitors: a review. J Clin Psychiat 44: 280–288

Tyrer P, Casey P, Gall J (1983) Relationship between neurosis and personality disorder. Br J Psychiat 142: 404–408

Walsh BT, Stewart JW, Roose SP, Gladis M, Glassman AH (1984) Treatment of bulaemia with phenelzine. A double-blind, placebo-controlled study. Arch Gen Psychiat 41: 1105–1109

White K, Simpson G (1984) The combined use of MAOIs and tricyclics. J Clin Psychiat 45: 67–69

Young MA, Sheftner WA, Klerman GL, Andreasen NC, Hirshfeld RMA (1986) The endogenous sub-type of depression: a study of its internal construct validity. Br J Psychiat 148: 257–267

Zahn TP, Insel TR, Murphy DL (1984) Psychophysiological changes during pharmacological treatment of patients with obsessive-compulsive disorder. Br J Psychiat 145: 39–44

Zametkin A, Rapoport JL, Murphy DL, Linnoila M, Ismond D (1985) Treatment of hyperactive children with monoamine oxidase inhibitors. Arch Gen Psychiat 42: 962–966

Authors' address: Dr. J. H. Dowson, Department of Psychiatry, University of Cambridge Clinical School, Addenbrooke's Hospital, Hills Road, Cambridge CB2 2QQ, United Kingdom.